LAND AT WAR

The official story of British farming 1939-1944

1/6
NET

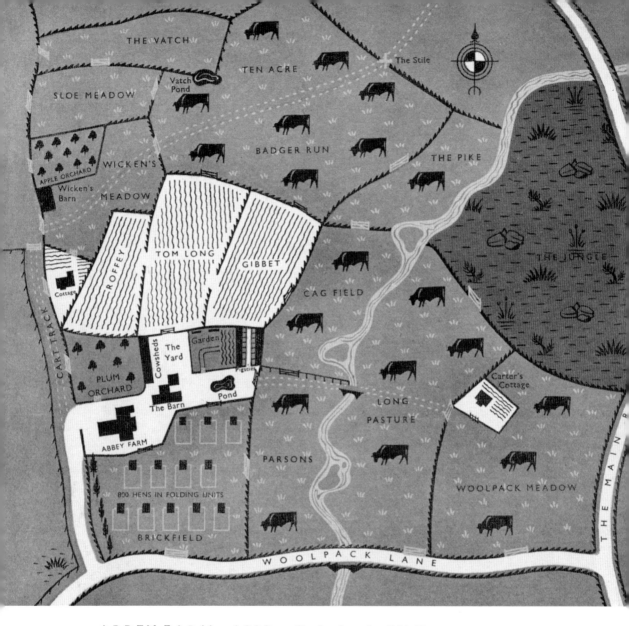

ABBEY FARM: 1939 *Predominantly Old Pasture*

(The facts and geography shown in these drawings are based, with slight modifications, on an actual farm in the South of England. So far as any one farm can be said to be typical, it indicates the general direction in which mixed farming has moved during the war.)

SIZE 100 acres (15 acres arable, 85 acres grass).

STOCK 20 to 40 Devon bullocks, fattening on pasture and imported cake.
800 hens in folding units, fed mainly on imported corn.
350 store pigs in yards, fattened mainly on imported meal.
2 horses.
Geese, turkeys, and occasional folds of sheep.

IMPLEMENTS 1 horse-plough, 1 horse-drawn grass-mower, horse-drawn cultivators, and reaper.

LABOUR 2 men

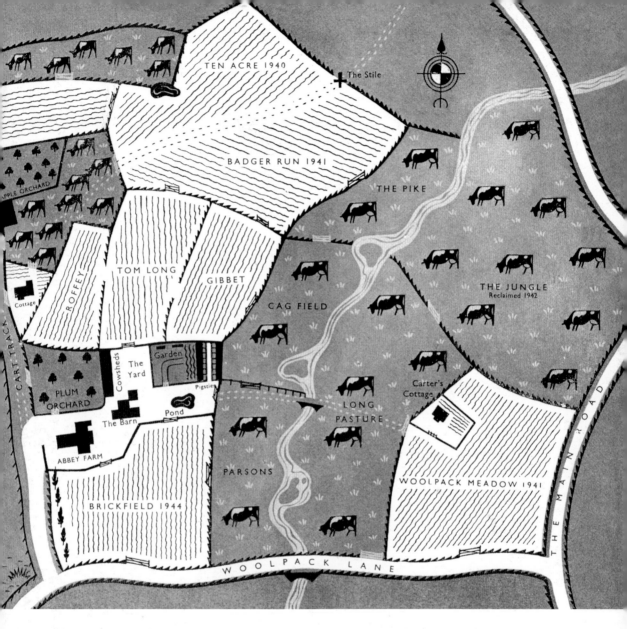

ABBEY FARM: 1944 — Arable balanced with Grass

SIZE 116 acres. 58 acres arable, 58 acres grass (16 acres added by reclamation).
STOCK 20 milking cows, 15 young stock, 1 bull, 25 pedigree sows and boar, 150 young pigs, 1 horse.
IMPLEMENTS 1 tractor, pulling plough, cultivators, 3-row potato ridger and reaper.
LABOUR 1 man, 1 land girl.

WAR CROPPING	1940	1941	1942	1943	1944
ROFFEY	Wheat	Clover	Oats	Beans	Oats, and 3-year ley
TOM LONG	Potatoes	Wheat	Oats	Roots	Oats, and 3-year ley
GIBBET	Wheat	Linseed	Potatoes	Spring Barley	Clover
TEN ACRE	Winter Oats	Winter Wheat	Clover	Potatoes	Spring Barley under-sown with Clover
BADGER RUN	Grass	Winter Oats	Winter Wheat	Clover	Spring Potatoes followed by Roots and Kale
WOOLPACK	Grass	Ploughed and slagged	Spring Oats and Peas	Winter Oats	Potatoes
SLOE MEADOW	Grass	Grass	Grass	Winter Wheat	Oats and Beans
BRICKFIELD	Grass	Grass	Grass	Grass	Spring Linseed
THE JUNGLE	Briars	Briars	Reclaimed	Re-seeded	Pasture

Land at War

prepared by

the Ministry of Information

London: His Majesty's Stationery Office

SEED MIXING

PLOUGHING

GROOMING

CONT

1   *War comes to the land*   PAGE   7

2   *The return of the plough*   PAGE   15

3   *Factory on wheels*   PAGE   22

4   *New harvests and new men*   PAGE   30

5   *Reclaiming the bad lands*   PAGE   38

CROWN COPYRIGHT RESERVED

Price 1s. 6d. net from His Majesty's Stationery Office at York House, Kingsway, London, W.C.2 ; 13a, Castle Street, Edinburgh, 2 ; 39-41, King Street, Manchester, 2 ; 1, St. Andrew's Crescent, Cardiff ; 80, Chichester Street, Belfast ; or at any bookseller.

Printed in Great Britain by THE WHITEFRIARS PRESS LTD.,

E N T S

6	The countryman's myriad foes	PAGE 49
7	Green pastures	PAGE 56
8	Farming in Scotland	PAGE 65
9	Ulster: a country of small holdings	PAGE 78
10	New life on the land	PAGE 85

FIRST PUBLISHED 1945

This book is prepared by the Ministry of Information with the approval of the Ministry of Agriculture and Fisheries, the Department of Agriculture for Scotland and the Northern Ireland Ministry of Agriculture.

London and Tonbridge.   S.O. Code No. 70-478*

SHEARING

THATCHING

MILKING

1

War comes to the land

NO WEAPON ever invented is more deadly than hunger; it can spike guns, destroy courage, and break the will of the most resolute peoples. The finest armies in the world, courageous enough in the face of bombs or bullets, can be reduced by it to helplessness and surrender.

This is the story of Britain's battlefield—the land; and of how 300,000 farms, strong-points in the battle against hunger, were armed, equipped and manned, so that the rich though neglected soil of these islands could be won back to fertility and help to feed and sustain a nation at war.

When, in 1939, we turned again to the land, we found it no more prepared for war than we were ourselves. It is true we had done very little to keep it trimmed for such an emergency. The Government had passed a number of measures to help the farmer, and the Agricultural Departments had worked hard to keep him on his feet, but 20 years of easy-going peace made that task extremely difficult, and the state of the world in general had, in fact, done British agriculture little good at all. The hungry lessons of the last Great War were too soon forgotten; we made promises to the farmer, but few survived for more than a couple of years. A large number of farms which in 1917 had only just saved us from starvation were allowed to slip back from cultivation to ranching, a process of neglect and deterioration began, and it was not a happy story.

There are few farmlands in the world more lush, more responsive, or able to bear a greater variety of crops than those in these islands; there are few men with a better natural feeling for the land than that which our farmers and farm workers possess. They are the sons of Britain's oldest industry, the inheritors of a complex and highly developed craft. With their eye for animals, their love of the soil, their capacity for hard work and endless experimentation, their genius for cattle-breeding, and their prophetic sense of the weather, they are the product of centuries of natural wisdom.

Little of this can be learnt from a book; it can only mature slowly in the brains of a long succession of land-workers who have proved by their experience the reality of what may seem to us mere guess-work or superstition, but is in fact scientific truth. Because of their long centuries of achievement British farmers had a name in the world; they had bred such excellent strains of cattle, wheat and grasses, that many younger agricultural industries abroad were founded almost completely upon them. Agriculture, with seafaring, has always been one of the natural callings supremely important to an island people. We were never foolish enough to lose our mastery of the sea;

it seems all the more incredible that we should have forgotten our need for the land and allowed so much of it to deteriorate between the wars.

There were still, of course, many farmers in this country who were both modern and progressive, and were running their farms in the best possible traditions. In spite of the hard times, these farms endured by the merits of the men who ran them, and were to form a valuable advance guard, when war came, in leading the whole industry forward into war production. There were times, however, when they appeared rather as islands of good fortune in the general apathy from which the countryside was suffering.

Before the war, of the 48 million people packed into these islands, only about one million families were engaged directly in the production of food. Of the remainder most were concentrated in the cities. It is obvious that, with such a vast urban population to feed, Britain could never be wholly self-supporting, and the existence of those dependent urban millions constituted one of our gravest national problems. Influenced by the readiness of foreign countries to provide us with cheap food, we allowed our home production to fall far below the margin of safety. Prices dropped, and many a farmer, with little encouragement to grow crops except on the best land, concerned himself chiefly with cutting down expenses by turning to livestock farming on the ranching system. In the worst instances the fields became little more than exercising yards for this stock. Fed almost wholly upon imported food and owing little to the land they walked on, cattle were merely incidental to a process which turned the raw material from abroad into milk and meat. Many farmers were forgetting the use of the plough altogether; they were becoming no longer cultivators of land, but cattle-ranchers.

Black as this picture is, it does not, of course, relate to the whole of British agriculture; but it does express something of a disease which was world-wide and had begun to strike seriously at the weakest links of our home industry. It was a problem with which the Agricultural Departments had been wrestling for some years. Much had been done to alleviate it. Subsidies and import quotas, and the vast marketing machinery built up from 1930 onwards, were by no means a wasted effort. During the decade which preceded the outbreak of war, these measures evolved and became the basis for a well-organised system of marketing and distribution of some of our agricultural products—a system which was more or less ready by the time war broke out. Efforts had been made, too, to build up fertility, and to accumulate a reserve of machinery for a time which some felt was bound to come. But because of the general apathy on the part of the layman these efforts, valuable as they were, had not achieved that degree of readiness which agriculturists would have liked, and when war did come, the Agricultural Departments were still faced with many serious problems.

This, then, was the situation—Britain depending for over 60 per cent. of her foodstuffs from overseas sources, and her supply lines in danger of being cut at any moment; a vast population of soldiers and factory-workers to be fed; and much of the land out of heart and in a worse condition than it had been for many years. How, then, to avoid starvation? How, with too few tools, and with labour short, to plough the land, feed it, and cajole it back into good humour so that once again it might bear us ample crops? That was the problem.

It will be difficult to forget the strangeness of that first September: the breathless days of Indian Summer pregnant with news of war, the heat haze over the yellow fields, and the silence of earth and sky. The countryside was not to know that silence again for some years: gradually those skies began to shake with our heavy bombers, huge tanks would squeal through the narrow

" . . . *with their love of the soil, their eye for animals, their capacity for hard work* . . . "

BAD TIMES. Twenty years of peace were not kind to every farmer. Tumbledown buildings, idle tools, fields strangled by weed and flood—these were legacies of ill luck, bad markets, slump. This had to be made good.

dusty lanes, and the fields themselves become a day-long clatter of drill and tractor, the tools of the land's offensive.

But in those early days there was none of that; the land lay listening, waiting for something to happen. It did not wait long. Already German submarines were out in the Atlantic, nosing across our sea-routes; and the explosions which shattered the hull of the *Athenia* came reverberating through the summer air, over the fields, up and down the quiet valleys like a note of warning. The blockade had begun.

Once more, then, Britain was alone, an island in the North Atlantic, packed with millions of people subject to the peril of starvation. No one knew just how real that peril might be. Suddenly all the overseas foodstuffs on which we relied became uncertain. We no longer knew whether we should get any at all. And once again we turned to the farmer for salvation.

The immediate job was to get as much land as possible ploughed up and planted by the following spring. The national larder was threatened, and thousands of tons of corn, potatoes and cattle fodder had to be found to fill it. There was not a moment to lose, for the need was urgent, the year was growing old, the land had to be ploughed and winter corn to be sown before it was too late. Every farmer was waiting to enter the battle, to be told his station, and put to the test. Not a moment was lost, for the Agricultural Departments had their war plans ready and brought them into action right away.

To start with, the Government needed leaders, men of example and imagination, men who had dirt on their boots, who knew the land; men, moreover, who spoke the tongue of the farmer, who knew his life and problems, and in just what way to ask for the impossible thing. For much that was formerly impossible had now to be done. Fortunately such men were not hard to find, and all over the British Isles the toughest and most practical farm men in every county had already been chosen and formed into Committees to organise and inspire the work of their particular area. At the same time a Women's Land Army Committee was appointed for each county and empowered to find local representatives, whose job it would be to watch over the welfare of the land girls coming into their district.

On Sunday, September 3rd, these War Committees were called by telegram to immediate action, up and down the country. They were empowered by the Minister of Agriculture to enforce his special wartime measures. Yet theirs was a form of self-government, and their brother farmers, knowing that these powers would be exercised not by some band of remote officials but with the sympathy and understanding of their own kind, were the more content and the more willingly co-operative. Farm workers' representatives, who also sat on these Committees, gave helpful advice on labour matters and encouraged the workers in their efforts.

The Committees set to work without delay. They met in village inns, barns, town halls and farmhouse kitchens. They kept their talk short, for the time was short, and they had to get back to their fields. But each knew he was in the battle at last, and when he returned to his farm that night he had all the responsibility of his district upon him, and a list in his hand which said: " 10,000 acres wheat, 6,000 acres potatoes, 1,000 acres sugar-beet, etc.", or more, or less, according to the size of his district. This was his immediate objective, and as he looked round the local landscape of unploughed grass, he may well have wondered if next spring would really see so many things growing, for no other spring had done so in 20 years. But that night or the following day he went round among his neighbours and showed them the list, and when they had all finished laughing, one or the other would say: " Of course it *might* be done, but I'm hanged if I can see how ".

DOMESDAY BOOK 1066, in which William the Conqueror reviewed the wealth of his new conquest. Above are listed the ploughs, cattle, "hides" of land, which he found at Hendon.

Later, District Committees were formed with, if possible, a representative from every parish, and they put their heads together and in the end every local farm and field knew how best it could play its part.

That is how the somewhat prosaically termed County War Agricultural Executive Committees came into being. But do not let that word "Committee" mislead you. Here was no talk-shop, but a hard-bitten band of fighters who had a very real and critical battle on hand. Their first job was to bring in the straggling peacetime harvest as quickly as possible, to clear the tangled fields for action, and get two million extra acres of land ploughed up and under crops by the following year. This they achieved by what is perhaps the most successful example of decentralisation and the most democratic use of "control" this war has produced.

From Whitehall to every farm in the country the C.W.A.E.C.s formed a visible human chain, a chain which grew stronger with each year of war. Here, roughly, is the way it worked. The Government might say to the Minister of Agriculture: "We need so much home-grown food next year". The Minister assured himself that the labour, tractors, equipment, and so on, would be forthcoming, and said to the Chairman of a County Committee: "We've got to plough two million extra acres next year. The quota for your county is 40,000".

The Chairman said to his District Committee Chairman: "You've been scheduled for 5,000 acres".

The Committee-man said to his Parish Representative: "You've got to find 800 acres, then".

And the Parish Representative, who knew every yard of the valley, went to the farmer at the end of the lane.

"Bob," he said, "how about that 17-acre field—for wheat?"

And Farmer Bob said "Aye".

For the C.W.A.E.C.s, instead of issuing orders from the remote anonymity of a Whitehall desk, went out into the fields, into the barns and cowsheds, into the pubs and market-places, and talked, argued and pleaded with their fellow-farmers to produce what was needed. The result more than justified these methods of peaceful persuasion, and was a testimony to the public spirit and adaptability of the British farmer.

Thus the Agricultural Committees went to war, and with them a tough but independent army of farmers and workers, an army that had not only been through a bad time, but were then very short of the weapons for the difficult battles facing them. It was the task of the Committees to inspire that army, equip them, and show them what must be done. The old, independent peacetime methods of farming would not do any more. The job on hand demanded completely new methods, modern methods, such as some farmers had not even heard of.

To help the Committees, the Government provided them with a staff of experts, under an Executive Officer, to spread these modern methods, to explain and popularise them. New personalities began to appear in the countryside; there were the Cultivations Officer, the Technical Officer, experts on silage, straw-pulping, ley-farming, farm-drainage, milk production, machinery, fertilisers, pests, and plant and animal diseases. Apart from the voluntary, unpaid Committee-men who were the prophets of the new farming, and who gave up all their spare time to the job, the Ministry of Agriculture roped in the best technical brains in the country: scientists, specialists, and young men whose studies of some particular branch of husbandry had taken them all over the world, to supplement its overworked nucleus of experts. It was the farmer's job to produce the goods, but he now had at his disposal the free advice and assistance of these experts on every conceivable problem.

With so much to be done, we had to have a full knowledge of our resources. Later on, a National Farm Survey was begun—a second Domesday Book—to record the state of every farm in the country. Nothing like this had been attempted since the eleventh century, when William the Conqueror set out to discover the wealth of his new conquest.

"*Then sent he his men all over England into each shire commissioning them to find out how many hundreds of hides were in the shire, what land the King himself had and what stock was upon the land, or what dues he ought to have by the year from the shire.*"

But the aim of this present survey was somewhat different. Agriculture at war had to know the exact strength of every fighting unit. Hundreds of field-workers, mostly volunteer Committee-men or retired farmers, began the gigantic task of surveying every holding with more than five acres of land. They covered every shire and parish; they

B. CONDITIONS OF FARM.

	Heavy	Medium	Light	Peaty
1. Proportion (%) of area on which soil is		45	25	
2. Is farm conveniently laid out? Yes ...				X
Moderately				
No ...				

	Good	Fair	Bad
3. Proportion (%) of farm which is naturally ...	65	35	
4. Situation in regard to road	X		
5. Situation in regard to railway		X	
6. Condition of farmhouse ...	X		
Condition of buildings ...	X		
7. Condition of farm roads ...		X	
8. Condition of fences ...		X	
9. Condition of ditches ...	X		
10. General condition of field drainage		X	
11. Condition of cottages ...		X	

	No.
12. Number of cottages within farm area ...	2
Number of cottages elsewhere ...	0
13. Number of cottages let on service tenancy ...	1

FARM SURVEY 1940. Nine hundred years later, as a new invasion threatened, Britain made a second survey. Every farm was noted, with the state of its buildings, roads, fences, and its soils.

worked with 6-in. scale ordnance maps, tact, circumspection, and plain physical stamina. For they not only had to assess the qualities of the land, they had to sum up the qualities of the farmer himself.

In the end, they had searched thousands of square miles of country, recording in detail the condition of each farm, the state of the land, the types of soil to be found there, the acreages of crops, acreages of grass, and the areas of dereliction. They had noted, too, the state of buildings, cottages, cart-roads, fences, ditches, drains, water and electricity supplies; the degree of infestation from rats, rabbits and other pests; and whether or not the farmer was a good one.

Britain is not so well known as we imagine it; main roads have worn a familiar track across particular stretches of country, but away from these you are often in lands and valleys unfrequented, and unchanged by the years. All these were explored at last, many strange facts discovered, and many old mysteries cleared up. Grazing lands, held in trust by villages, about which no documents existed nor any proof but the proof of tradition, were noted down for the first time in centuries.

But particularly the good farms and the bad farms were noted, and the reasons for the latter recorded—lack of adequate roads, lack of lime, etc.—so that steps might be taken to improve them. The job took a long time, but it was not attempted out of mere curiosity. There now exist detailed maps upon which, outlined in different colours, 300,000 farms are marked and known. This Farm Survey may well eclipse the Domesday Book not merely in comprehension but in historical importance : it not only provided invaluable data for the wartime mobilisation of our resources, it has helped to establish a blue-print for post-war agricultural planning.

But long before this Survey started, the great ploughing-up offensive had begun. County Committees, having allotted targets to every farm, began to organise labour for those who needed it. And the farmers, clearing their peacetime harvests, or those without harvests at all, began to look to the approaching winter of 1939-40 as a time for unprecedented activity. Two million acres of old grassland had to be turned over in a few months, had to be turned face downwards so that frost could work upon the weeds and grubs, could break up the soil, clean and prepare it for the first wartime crops upon which so much depended. It was a bitter winter, harsh and uncompromising, but for the farmers of Britain it marked their first victory.

PLAN OF OPERATIONS. Farming is a battle against time and weather. Crops must be planned, labour and machines organised, seed sown at exactly the right moment. The land does not wait.

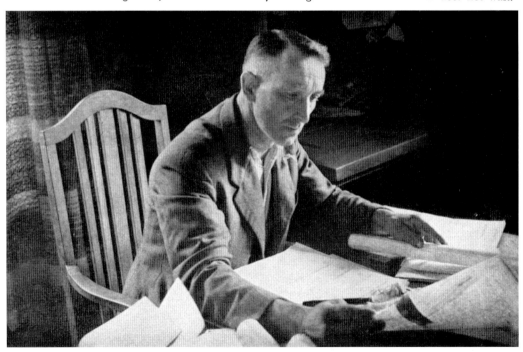

2

The return of the plough

THOUGH the sword has long been a symbol of war, the ploughshare one of peace, to-day this symbolism is no longer true. For in the hands of the modern British farmer the plough became simply another weapon of war, playing its own specialised part among the deadlier armament to which the sword itself had given place. Invention complicates both life and battle, but the physical needs of a besieged island remain very much the same, and old problems recur in each new moment of crisis.

It is a long time since the Spanish Armada swept with so much bluff and bullheadedness up the English Channel; but for us, in 1939, the situation was very similar. As flaming beacons greeted the Armada and spread their warning from hill to hill, rusty swords were taken down from the walls and polished for battle. Queen Elizabeth, gazing upon the quiet sheepwalks of the countryside, knew the danger in their idleness and persuaded the farmer, by forcible methods, to bring out his plough and look to his land again.

In 1939 the threat of starvation was just as real, but our own farmers, unlike those of Elizabeth's day, were keenly aware of the country's need, and of the necessity for getting more land under cultivation. The Government was offering a small grant towards the cost of ploughing each new acre of land, but this, of course, was no more than a token to the farmer. For a sudden return to the plough cannot be undertaken without at least temporary hardship. Switching from the regular profits of milk production to the long-term process of arable cultivation upsets for a time the whole economy of the farm; it means hard months of labour with no immediate reward, and always the risk of crop failure or some such disaster.

Yes, there was more to it than the paying of subsidies. You can't plough a field with a pound note. Thousands of skilled workers had been taken into the army, the weather was bad, ploughs were scarce, horses scarcer; nor were there nearly enough tractors. But the job offered such vast opportunities that even the most serious difficulties were overcome. There were large areas of permanent grass and rough grazing in England which the plough had not visited for years. Much of Wales was in a like state, though wilder, her mountains and valleys given over almost entirely to sheep and cattle breeding. Even among the best farms there were fields it had not paid to cultivate till now.

It was against these millions of ripe but ragged acres that the plough-up offensive was directed. Dunged by generations of cattle, enriched by years of rotting and undisturbed turf, fertility lay like a priceless vein beneath the surface, waiting only for the plough's knife and the farmer's skill to

THE PEACETIME STUBBLE, and the untouched grass, each was ploughed in turn, ready for the first crops of war. And wherever the teams went out the gulls followed, foraging in the open furrows.

release it and turn it into bread. Hardly ever before in his history had the British farmer been faced with such a task or such an opportunity; it was gigantic both in difficulty and promise. But the need of the country created the will to ignore difficulties. They were nothing compared with the urge to get the land opened up.

Every implement, in no matter what condition, became a treasure without price. Old ploughs were brought out, oiled and renovated. In a shower of sparks and clamour the local repairer and the village blacksmith came into their own again, sharpening plough-points, linking up chain harrows, refitting broken shafts, and forging new blades for the struggle. Old harness was cleaned and cobbled; horses teamed up; and the garages of the small towns began to attack the novel problems of tractor maintenance.

Early in the autumn of that first year the general offensive began. The first teams looked rather small setting out over those miles of grass; the furrows they left looked slight compared with all that had to be done. But slowly the countryside began to change; the dry yellow old grass sank beneath the waves of heavy loam thrown up in field after field. Green pastures broke into vivid patches of vari-coloured earth which grew and spread and joined with each other. The land began to assume a new fabric, a texture and richness much of it had almost forgotten. And the old men would look out across the fields and say: "They bin and ploughed the meadow, look. Ain't seen that done since the last war. Well I never".

All this happened in its own time, in the time it took to share out the tools available. The farmer might have to wait, impatient at the passing of a bright day. Then his neighbour would arrive offering his plough, his horse, or his tractor. Or teams of contractors would come hot-foot from another job, the men hard-driven, in demand everywhere. In the first few weeks it was a case of each doing his best alone, or getting help wherever he could. Soon, however, things became easier and the County Committees were organising their own forms of contracting service. A nucleus of machinery was put into their hands and this they sent out with its own labour crews wherever it was needed, tackling the work of cultivation in districts where farmers were short of equipment or where contractors could not cope with the task. Later they arranged to hire the necessary implements to farmers and contractors direct.

As the campaign developed, Committee offices came to be like military headquarters. Wall maps, dotted with a multitude of flags and coloured pins, showed the position of machinery depots, the movements of mobile units. The telephone rang continuously—accidents, breakdowns: a man in trouble, sick, unable to reach his objective; the nearest unit would be moved in to his aid, reinforcements sent elsewhere. To save time, and to prevent the overlapping of journeys, a complex strategy had to be observed. Time, and the most effective placing of equipment, were of supreme importance. Tractor drivers worked till they were fit to drop, land girls shared their hardships, and the farmer spared neither them nor himself.

But in this first season of 1939-40, before the W.A.E.C.s had fully organised their machinery depots, it was a question of every man for himself. Horses and men had to break off sometimes and rest, but an idle tractor in those days was an abomination no one would stomach. Wherever possible they were manned in shifts, and at mealtimes relief drivers took over when such were available.

Early in 1940 there came prolonged frosts which held up all work on the land for weeks at a stretch. It looked as if the job would never be finished; but when at last the frosts lifted, ploughing began again with a vengeance, the men worked seven days a week, all day and sometimes even in the dark.

THEY PLOUGHED THE HILLS AND VALLEYS, THEY PLOUGHED TO THE VERY EDGE OF THE SEA.

THEY PLOUGHED BY DAY AND BY NIGHT, THROUGH ONE OF THE HARDEST WINTERS IN LIVING MEMORY.

ON THE SOUTH DOWNS, unfarmed since the days of the Saxons, the modern plough appeared.

Night ploughing, though never a very common practice, developed, when necessary, its own special technique. It had to; for, even in daylight, ploughing is a tricky business; you need a clear and confident eye for it. But to plough at night is like a blind man walking a tightrope; shadows deceive you at every turn, you lose your sense of direction, and to control your machine and keep your furrow straight you need the instincts of both owl and acrobat. But experienced ploughmen soon developed a sixth sense; they got used to ploughing by moonlight, and when there was no moon they worked to a special system of lights: one fixed low on the front of the tractor and screened to light up the furrow, and another—a lantern—hung in the hedge before them and moved after each cut to give them direction.

But it was a strange business; it had that quality of eerie beauty which so much nocturnal wartime activity possesses. Like searchlights, flarepaths, the floating stars of night-flying aircraft, these lighted tractors were but part of the new pattern. Crawling along the sides of a hill, their lamps shining through the bare trees, they looked like glow-worms moving through grass. The muffled sound of their engines, rousing the winter darkness which would otherwise have been so still, brought home to the countryman the restless urgency of those days.

From coast to coast now the ploughs were fanning out over new ground. They were climbing the Lower Pennines, breaking the Devon moors, sweeping long brown paths over the Berkshire Downs, and threading their way between the derelict bungalows of Sussex holiday camps.

The old skill in horse-ploughing still played an important part; but a new skill was being born, and the tractor-plough, drawing two, three, even five furrows, began to show more than ever before its remarkable power and speed. For one tractor does the work of many horses and men. Fitted with caterpillar tracks, it can climb hillsides hitherto unploughable, and drawing the heavy blades of special ploughs can tackle rough ground no horse would ever look at. Yet the horse and the tractor are not enemies. To the delight of some of the older farmhands, the horse can still do things impossible to the tractor; an awkward corner, a tricky gradient or hollow—here it is still most effective.

For the men who worked with horse or tractor through that first winter it was no easy life. Like their brothers in the factory, they often did seven days a week for weeks unbroken. They worked through the most bitter weather, by daylight, moonlight and lamplight. They worked exposed to one of the most severe winters within living memory, and were often held up by ground which seemed as hard as rock. Horses would sometimes be found with their tails frozen to the stable floor, tractors would have to be thawed out with blowpipes, plough-points would snap off in the iron soil. Men were frost-bitten and snow-blinded; freak storms killed birds in mid-air, split open trees, and wrapped thick coats of ice round every twig and blade of grass, so that the slightest movement of the wind played them like a xylophone. Such weather was a grave hindrance to operations.

His life has hardened the land-worker to such trials; but for many hundreds of land girls fresh from the offices, the centrally heated shops, the soft life of the city, they were an ordeal which only enthusiasm could overcome. Even during the great thaw it was little better for them. For, whether you are stumbling behind a plough or perched high on a tractor, you cannot for a moment escape the temper of the elements. If you are walking on heavy land, you wear a ball and chain of mud which you drag wherever you go. And you have to scrape down the plough after each furrow; or the horses' hooves, or the tractor wheel. If you are on high land you might as well be at sea in an open boat; the wet winds come at you from every side, you are soaked with rain, while you feel the whole winter drain through your bones.

But, in spite of everything, the huge programme began to be achieved. By the spring of 1940 the job was done, the land lay ready for the first wartime sowing—two million new acres of it. Soil of many types and degrees of goodness. Boulder clay, Sussex chalk, Cotswold limestone, black Lincoln peat, Essex gravel, red Severn sandstone—all carved and cut and sliced and broken despite a savagely unfriendly season.

Looking upon those patchwork fields, new sign of the land's good heart, the farmer, his workers, not least the land girl, knew the cost and satisfaction of their accomplishment. The first line was taken, ready for consolidation. It was land the war made precious.

3

Factory on wheels

THE WINTER CORN survived; the spring wheat grew strong and lusty, spreading broad sheets of blue metallic leaf over the fresh ground. Here was a visible promise of the harvest to come, something real to justify the bitter months now past. But there was already more to be seen there than the mark of the plough. Other signs began to appear of even deeper significance—the sight of unfamiliar machines working the sprouting fields, cleaning, rolling and cultivating.

A hundred years ago power came into industry; the railroad replaced the packhorse, the steel foundry the blacksmith, mass-production the craft of the individual. Smoking cities spread out and dominated vast areas of country, drawing thousands of rural workers into the factories. Agriculture was left in a state of isolation by this upheaval, cut off from the industrial community by an ever-widening gulf of different traditions and experience. The application of power to the land was to bridge that gulf, was to compensate for the drift of men from the land, attract fresh blood to it; was to increase both the scope and tempo of agriculture and bring it into line with other forms of human activity—particularly war.

This change had already begun to manifest itself; war merely speeded up its development. The old inherited principles of good husbandry remained, but mechanisation was giving them a special twist. The whole direction of British farming was changing, not only in its capacity to work more land with less labour, but in the type of crops to be grown, the type of cattle to be raised, even in the methods of feeding them.

The problems of intensive cultivation have always been labour and time. Crops have to be planted at the right moment, or they may never flourish. And where the climate is as unpredictable as it is in Britain, the right moment has to be seized and exploited to the full, for it may not come again. Even if there is an abundance of physical labour, each man has only so much strength, and the day is only so long.

But the time had come to ignore such limitations. Complex and critical operations had to be carried out in a very short time, vast areas planted, worked and harvested. The farmer was faced with a severe test; and it was the machine that helped to solve problems both of time and accomplishment by extending enormously man's power over land.

By 1939 the mechanisation of agriculture had already reached a high stage of development abroad. The enormous farmlands of America, Canada, Australia, and the Soviet Union, presented problems of space which could never have been solved without the

TRACTORS ON PARADE. New from a British factory, they will give the farmer power for ploughing, sowing, reaping and threshing. They lead the mechanised regiment of agriculture.

use of great technical skill and imagination. Powerful and ingenious machines had long been in use there, designed to cope with all the complicated processes of cultivation in fields often larger than an English parish. Millions of fertile acres in the Ukraine, the Canadian wheat-belt, the American Middle West, would never have been cultivated at all without their aid.

To Britain, on the other hand, mechanisation came comparatively late, though its principles were already well established by the time war broke out and big strides have been made since then. By 1944 this country had over 175,000 tractors, compared with 55,000 in 1939. After the first five years of war Britain possessed one of the most highly mechanised agricultures in the world.

This was due largely to two things—the efforts of our own home industries and the generosity of our friends and allies. Inspired by the country's need the numbers of British-made tractors and other farm implements increased by leaps and bounds. Raw materials were scarce and the programme of war gave priority to the production of tanks, guns and other munitions. Labour was short, and skilled engineers practically unobtainable. Struggling continually to get the right materials and to maintain their labour force, the factories did a good job—far better than could ever have been expected in the circumstances.

The material help given by America and the Dominions was tremendous. Throughout the U-boat campaigns of 1941, and later,

of carnival to the dusty wagons upon which they rode. Many of them, unfamiliar to this country, were first taken and put through rigorous tests—a new type of plough, the prairie-buster, was tried out on a mountain-top in Wales: they were then distributed carefully among the farmers, the dealers, and the machinery depots of the County Committees.

Take a look at a typical Committee depot. Its home is a branch-line engine-shed in a small market town. The cindered yard looks as if a tank battle had been fought there; it is a mess of black mud torn up by passing tractors. The long shed contains a whole family of machines of all shapes and sizes, some draped with tarpaulins, some stripped for action; a mass of chains, sprockets, cogs,

PERIOD PIECE. The chattering saw-tooth blades of the old horse-drawn mower is a sound that has not yet died from our summer fields.

large consignments of machinery came over the seas. They came from Canada under Mutual Aid, from Australia also, and were among the Lend-Lease cargoes sent from the United States. They were shipped wherever there was a spare foot of space, on deck, or crated down in the deep holds with the heavy armament of tanks and guns; the space was not begrudged them. For these tools were to play an important part in the Battle of the Atlantic; they were to continue that battle on the soil of this country. Carried from port and factory, dazzling the eyes in their multi-coloured paint, these ingenious, complicated contraptions must have intrigued the ordinary man. For there is nothing so gay and gaudy, so fantastically shaped, so unlike anything else in the world as a piece of new farming equipment straight from the factory. But to the farmer who knew their uses they were the answer to many riddles, the means to accomplish the hitherto impossible.

No time was wasted in putting them to good account. They were sent out by rail to all parts of the country, lending a jaunty air

shares, spars, hooks, blades, revolving rakes and discs. Some look like early flying machines, some are as sleek as racing cars.

It is early morning. Mechanics are adjusting a seed-drill; they are skilled men, and their job is a full-time one, for the land gives hard wear to the toughest machine. Spare parts of all kinds are stacked near by; parts for tractors, ploughs, harrows, binders, threshers; of all shapes and sizes, from wheels and lengths of caterpillar track to plough-points and nuts and bolts. All are carefully labelled, for many are irreplaceable essentials without which the most costly machine would be of no more use than a piece of broken fence.

A group of land girls arrive, laughing and chattering. They start up their tractors, mount them and drive away, one after the other, each with a train of rattling equipment hooked on behind. One is going out to the lower hills to plant potatoes. Two others, dressed in oil-black overalls and mounted on a strange-looking machine with crawler tracks, are bound for the wet valley to work on a drainage job. This machine is a Cub excavator; it has a terrific, lurching power and a long snout armed with jaws and teeth of steel. It is astonishing that a machine of such size could be operated by girls at all. But these you see are chosen for their mechanical aptitude and given a special course of training by arrangement with the manufacturers. This excavator is used for digging and

COMBINED OPERATIONS. Cutting through the wheat in broad sixteen-foot swathes, threshing and bagging the grain in one continuous movement, the combine harvester makes short work of harvest.

TEETH OF THE GYROTILLER.

CUT OF THE PLOUGH.

cleaning field ditches; it can replace a whole gang of men, its jaw scooping out deep canals with speed and precision, and preparing the way for the work of other ingenious drainage tools such as the mechanical trencher and the mole plough.

This mole drainer is a subterranean tool. It is fitted with a heavy blade and a torpedo-shaped piece of steel—the " mole "—which is used to force a series of channels beneath the ground, linking the field with the piped mains which in turn carry the water into the ditches. The operation of mole-draining calls for enormous pulling power, which only the heavy crawler tractor can exert. Drainage work is some of the toughest farm work there is, and though some girls can tackle it, it is still largely a job for men.

Among the other heavy machines standing about the yard is the bulldozer, a battering ram with astonishing tenacity and strength. Mounted on **caterpillar tracks**, it thrusts

everything before it, slicing through hillocks, clearing the most unmanageable debris, pushing over trees. Another is the gyrotiller, huge, lumbering and slow, almost without equal in taming wild land. It combs the ground with thick steel prongs and stirs it up with a set of wicked-looking blades that revolve like an egg-whisk. This formidable combination can root out the toughest bushes and tear through the most obstinate tangle of roots and stones. This particular one has been working on the clearing of an old forest, its stubborn juggernaut progress combing out tree roots and matted briars at the rate of an acre a day.

The morning is still early and hardly light; the weak spring sun shines dimly in the yellow puddles, but the depot is swiftly emptying. Farmers, farm workers, Committee drivers, land girls, all wrapped in mud-caked mackintoshes, make a loud noise compounded of argument, advice and warning, as they

RAKE OF THE HARROW.

PRESS OF THE ROLLER.

collect their tools and drive them away down the chill echoing streets. As the year advances, this busy traffic of workers will continue, but the type of tool they ask for will change. Spring—like autumn—is a time for preparation, tillage and sowing, and the tools being taken out now are mostly ploughs and cultivators, disc harrows to slice up the furrow, combined seed-drills which sow seed and fertiliser together, rollers to cover and consolidate the seed bed, and the labour-saving potato-planter which is tuned to drop seed at given intervals.

The next demand will be for potato-ridgers, and machines with curious spasmodic movements designed for weeding, hoeing, or the transplanting of seedlings. Then the tools for harvesting : mowers, balers, rakes, elevators, for cutting, turning and stacking grass ; potato-lifters with revolving forks that throw up the tubers for the pickers to gather. Then the sturdy reapers and binders, veterans of long service but still capable of good work. And finally those giants of the fields, the combine harvesters—complex machines which, manned by one or two men, concentrate all the many processes of grain harvesting—reaping, threshing, drying and bagging—into one swift, continuous movement.

On big fields, where the grain is in good condition, these machines can clear as much as 20 acres a day. They are invaluable in bad weather ; by their speed they have saved thousands of tons of threatened wheat from otherwise certain ruin. One Cambridgeshire farmer, racing a storm, brought four of them to his fields as a last hope to save the ripened grain. He drove them through the wheat together in echelon, clearing 36 feet in a single sweep—compared with the five-foot cut of the old horse-drawn binder. When the storm broke, the grain was safe in the barn instead of soaking in the fields.

DRILL OF THE PLANTER.

ARMS OF THE REAPER.

That is mechanisation—a thing of a hundred complex, specialised tools and gadgets, automatic substitutes for the spade and the sickle, for the hand of the reaper, the hedger and ditcher, the milkmaid, muck-spreader, the turnip-hoer. In terms of gears, cranks, hooks and knives it is all that and more. It is a factory on wheels, a factory that passes over the land, preparing, tending, extracting food. And the tractor is its power unit, pushing, pulling, feeding it, serving its manifold processes. Surrounded by this range of equipment, the tractor is the king-pin of mechanisation, the heart of the modern farm. With these machines it runs and climbs, works the hills, and the fields; ploughs, sows, discs, rolls, reaps, threshes; carries the worker to his job and the gathered harvest to the mill.

Some of these tools were new to the farmer; some were improved versions of those he had been using for years; some had been developed from foreign types; but many were a direct result of his own invention and wartime need. Over two-thirds of the whole mass used in this country were produced in the war factories of Britain.

Farming was once thought to be a slow, back-aching business of hoe and hand-rake, and the farmer a poor benighted fellow, his craft the very antithesis of science and modern industry. But farming *is* a modern industry, its tempo is tuned to the machine, and the only thing that remains slow about it is the natural growth of the crop.

The job of the farm worker has changed too; he has put off the smock and put on the mechanic's overall, he works now with the tractor rather than the hand-tool; and his job has been the more exacting in skill since it demands not only an understanding of the machine, but still, as ever, an understanding of the land. The farmer and farm worker have had to become the masters of their new equipment. Their talent for improvisation and invention has always been highly developed; it had to be, for farming has a habit of throwing up problems which must be solved on the instant by intelligence and intuition. But in this war the landworkers not only improved and adapted by their own experience many machines coming into their hands: during the early shortage of tools, they often rigged up the most remarkable substitutes on their own.

So power came to the land, such power as it had never seen before—power to move mountains, drain marshes, to turn bogs into cornfields and cover hills with potatoes; power to fight weather, disease, thorns, rocks, wilderness; power to sow crops without hands and to harvest without loss; power to win food from the most difficult soils, to spread grass over bracken, to feed beasts where no beast could live before; power which could often be operated by girls and could do the work of armies.

Through it, agriculture found its feet again, stretched out, and tested its new strength. It will never again be just a country cousin; its life is ahead of it, equal to anything the modern world can bring. And, unlike those other symbols of power seen on the battlefield—contraptions bred only for war and death—the farm machine is designed for the living. It can be part of an expanding and highly developed industry, whose work in peace will be as vital to the hungry world as anything it did to help us win the war.

4

New harvests and new men

BY THE second year of war it was obvious to everyone that the countryside was astir in a big way. Never had the fields looked so well ordered, or the hedges so trim and well cared for. As more and more fields were made ready for planting, this bloom of recovery increased. On every hand there was an air of polish and purpose, a creaking and bustle of mounting activity; it was as if some vast empty mill had reopened its shuttered doors and was slowly returning to its original business.

Yet this was no one-man concern; it was made up of many individual units, not mixed farmers only, but shepherds, stockmen, fruit-farmers, seedsmen, pig-breeders—men who had developed a flair for a particular job and whose talents and experiences were widely varied. The fact that they could now gather as it were under one roof and work to a common programme says much for their adaptability. From many a farmer—as from many another type of craftsman, too—war required a complete change of task. He had to accustom himself to unfamiliar materials, struggle with new methods; he could no longer pick and choose his job. Crops that were new and untried to many, and food of all sorts in greater abundance than ever before—this meant upheaval and a new life to thousands of countrymen, particularly the older ones.

Somehow one expects the farmer to turn his hand to anything, but it is no easier for him than for any other man. To switch from pig-keeping to potatoes means sacrifice and the straining of new muscles; from flowers to wheat means a new trade and a new technique. Take the flower-farmers of Devon and Cornwall, for instance. On that warm limb of southern England they have always enjoyed a very fortunate climate. Summer stays late there, and spring comes early. Land of sub-tropical palms, of cream, wild strawberries and abundant flowers—here were grown the brighter things of life to supply the colder East and Midlands. But you cannot eat flowers. So the flower-farms changed. Mr. B. kept such a farm a few miles out of Helston. Once he grew 30 acres of daffodils and narcissi; he grew them from bulbs which had taken a score of years to refine and develop. In April 1940 he picked his last big crop and packed it off to London. He will tell you what happened after that.

"There was a time when you couldn't look at these fields without seeing daffodils, narcissi, anemones, tulips, flowers right down to the cliff edge, growing all through the summer. Well, there's a different sight to the land now, altogether. Not so much colour, perhaps, but a good deal more spice. I ploughed up my bulbs a couple of years

CROP FACE. *Above* are the silvered stems of flax plants, pulled by hand and laid out in the sun to dry. *Below*, a field of seeding carrots, growing where flowers once grew, their useful heads supplanting blooms of aster and chrysanthemum.

INTERCROPPING. Ploughing among fruit trees is a ticklish job, though good economy in war time. Fertilised by years of fallen fruit and leaves, the soil is rich and a heavy crop will grow here.

back. It wasn't a job I liked doing, I can tell you. Now I'm growing wheat, potatoes, carrots and onions. Some of my bulbs are still rotting in the ditch, others have been steamed and fed to the pigs. But I'm not wasting any sleep on them, that's all done with. We flower-growers are still allowed to keep on a small percentage, but I thought I'd make a clean sweep of it. This is good sheltered land, light and easy to work. Down by the sea I'm growing two crops a year, early potatoes followed by carrots. We've had help from the Committee with ploughing and lifting, but weeding's the very devil; the back of my neck is scorched to a cinder, and so is the missus's. Still we can't grumble; we got nearly a hundred sacks of grain last year, 50 tons of potatoes, and 20 of carrots. Land which grew good daffs can grow good vegetables, and I reckon it'll grow just as good daffs again if we want it to."

Flower-farmers all over the country were making a similar effort. And they were giving up a good deal. Acres of rose trees ripped up by the plough made room for wheat and barley; lilies and orchids cast out of the hot-houses were succeeded by more practical, less exotic plants — tomatoes, lettuce, mustard and cress. Those rainbow-coloured nursery fields, once a riot of carnations, lupins, roses and chrysanthemums, the despair and admiration of amateur gardeners, assumed an austere sobriety, spread with such matter-of-fact necessities as grass seed, potatoes, cabbage and parsnips. Ploughing in bulbs, rooting up rose trees and fields of valuable shrubs and plants—all this was no easy thing for the nurseryman to do. He gained nothing by it; it was often something of a real tragedy to him. He just knew it had to be done. But, with his acres of frames and glasshouses, and his unique experience

burning apple and cherry wood, and much good potash was added to soil already enriched by years of rotting fruit and leaves. Fields such as these produced bumper harvests.

Younger, healthier fruit-trees were spared, but the land between them was squeezed to the last drop of goodness. This sort of thing is called intercropping, and here is a Cambridgeshire man's account of it:

"I've 10 acres of apple, 10 of plum, and a dozen of cherry. The trees grow anything from 10 to 20 foot apart. For the past three years I've run a plough between 'em—rather a ticklish job till you get the hang of it. A tractor goes down the avenues and a horse-plough does the cross-work. In the old days those trees just grew, and there was always a lot of good grass going begging. Perhaps I'd run a flock of geese on it, or graze a few goats; I might even feed a pig or two on the windfalls, but I never tried intercropping

TRACTOR UNDER GLASS. Preparing ground for tomatoes in a three-quarter-acre greenhouse in the market-gardening country near Evesham.

of plant-rearing, he made his own valuable contribution, and often provided backyard gardeners with quantities of seeds and millions of young vegetable plants to bring on for their own domestic use.

With the fruit farms it was the same; technique and equipment were similarly adapted. In the orchards of Kent, Cambridge, and the West of England new crops sprang up between the severely regimented trees; no space was left idle. In some parts there were orchards, planted half a century ago, which had seen their best days. Crumbling old apple trees in advanced stages of decay, their warped trunks full of wood-lice and loaded with mistletoe and fungus, cluttered up first-class land. Such trees had to go, and many acres were cleared. Tractors pulled the trees down, and gyrotillers grubbed up their roots. Huge bonfires were built, filling the air with the aromatic smoke of before. Now I'm growing potatoes where those geese used to run; I lifted 50 tons last year and I reckon there's half as much again waiting to be lifted this. Besides which, I've fattened a score of heifers in the orchards and my fruit crop's as good as ever. I suppose it just goes to show the land we used to waste."

The plough brought new crops and a new outlook to another specialised type of farm —the stock-fattening grasslands of the South and Midlands. Beef is fattened on grass; and English grass, when it is good, can hardly be bettered for this purpose. Neither New Zealand nor South America at their best could ever show anything to compare with the deep lush pastures of Leicestershire, for example, or the Romney Marshes—at least, as they existed before the war. But such permanent grass and the beef it produces were expensive luxuries to a nation besieged; it could save more shipping by growing wheat and potatoes.

DOUBLE HARVEST. In a West-country orchard girls gather plums, while rows of inter-cropped potatoes flourish beneath the fruit trees.

The wartime story of the "Five Langtons" near Market Harborough is typical of the way in which many of our best grasslands were conscripted. What we saw there is perhaps the most significant and complete change-over to be experienced by any farmer. The parishes of Langton were unique. Their five church towers overlooked some of the best pastures in the world. Years of expert management had produced on those gentle hills grass of such astonishing richness that it was possible to fatten as many as two bullocks to the acre in a good summer. The soil was deep and the colour of milk-chocolate, the sward was like honey to the feeding cattle. It had been achieved by generations of careful grazing, weeding and dunging. Nobody dreamed of ploughing it; they would have thought it madness. As a local grazier put it:

"When you came to the Langtons you were in the Land of Goshen. Fields all covered with clovers, and ryegrass as rich as cake. We weren't farmers in this part of the world; we were graziers, and grass-management was our life. We could tell from the look of a field what would thrive on it—this for bullocks, that for heifers, the other for young calves—even the kids and the women could tell you. We used to buy in lean cattle from Ireland and Wales, strong Hereford beef bullocks and so on; turn 'em out to grass round about May, and just sit on our behinds and watch 'em grow. Fields were grazed down through June and July, then rested while the cattle went east to winter in yards. Aye, stock-fattening was a gentleman's life."

Then the time came when a much weightier yield was required from this land, far more than it would produce in meat. "It had taken some of us nigh on 50 years to get the grass in that condition. We thought the world had gone mad when they asked us to plough. We pretty near wept when it came to it."

But they ploughed. The fat-stock left the

land, and field after field came up, some for cereals, for potatoes, for sugar-beet. It was a hard new life. Travelling light with summer grass and cattle had been the graziers' existence: it left them with little over for farming proper; no arable experience, no equipment, little skill, few buildings—nothing but the land. Yet they found what was lacking. Skilled men from the eastern arable districts moved in to reacquaint them with the art of ploughing. They bought machinery and knocked up tools of their own. One of the first potato-drills to be seen in that district was rigged up by a local blacksmith. Soon, only a few of those fields carried the beef cattle of old; the rest bore a heavier harvest.

Old-time graziers saw these pastures go with some misgivings. In fact there was one occasion in the Romney Marshes when this feeling went so deep that an old man rose up and laid the " curse of the marshmen " upon a team of land girls who were ploughing the sacred turf. " Those pastures took 50 years to mature," he said; " they've gone and I shan't live to see them come back." He need not have worried. Scientific re-seeding can return those pastures to their former excellence, not in 50 years, but in as many months.

Cereals and potatoes were the main crops of war. If bread is the staff of life, the potato is its second prop; it was vital to us in war time, for it gives bulk, keeps well, and goes with anything. In 1917 it was only because we had a good stock of potatoes in hand that we survived the critical U-boat peril.

Nearly every farmer had to grow a quota of potatoes, and many new growers did not enjoy it. The potato is laborious, time-consuming, sensitive to disease, and hard to handle without special equipment. Nevertheless, each man did his share, and many lightened the task with various devices of their own. In the early days, when special equipment was difficult to come by, a Cots-

POTATO CLAMP. Warm under their blankets of straw and earth, they'll last the winter through.

wold sheep-farmer was asked by the local Committee to plant ten acres. He surveyed the heavy field with distaste; the prospect of sowing the potatoes by hand, eight inches apart, arduously, row by row, gave him no pleasure at all. The way in which he solved his problem has become one of the brighter legends of the village.

" I just happened to be going by his place one morning ", says a neighbour, " when I heard a noise like a gurt clock. I didn't know where it came from till I looked over the wall. Then I seen old Jesse. He'd got a three-furrow plough he was ridin' on and a tin bath full of spuds in his arms. He'd raked up an old chimney-pot from somewhere and got it wedged between his knees. There was some sort of gadget stuck on the wheel somewhere which rang a bell every time it turned round, and each time this

FRENCH STYLE CULTIVATION. Forcing lettuce under frames of glass breaks the monotony of winter cabbage by providing early salads.

bell rang old Jesse dropped a tater down the chimney. 'Jesse,' I said, 'that's as neat a contraption as I've seen anywhere'." Jesse was ploughing three furrows and planting at the same time, dropping the potato in the third furrow as he went along and so spacing out the rows. This was his own solution, thought up in his own bed. Many others conceived their own variations, no less ingenious.

Sugar-beet is another crop which increased and which many farmers were tackling for the first time. Before the war, the industry was subsidised, and many people were puzzled by this: sugar comes cheap enough from abroad, they said, why bother to grow it here? It is lucky for us that we did. The experience then gained by the growers was largely responsible for the great harvests gathered in war time. You could see the plants in many parts—scraggy tubers, looking no better nor any sweeter than common cattle-food. You saw it piled in railway wagons, or heaped by the roadside waiting for transport to the factories. You may not have been impressed by the sight of it, but that was your sugar ration. Very few ships were needed for sugar. Year after year the entire domestic ration of the country came from these muddy-coloured roots.

Yet another new name to the farmers is flax. Wherever the blue flowers grow they light up old fields like patches of summer sky or stretches of south-sea water. Cultivated hardly at all in Britain between the wars, flax became a crop of the first importance. It was utilised in the making of canvas, tents, maps, camouflage, aircraft fabric and parachute-harness. In Kent, Sussex and East Anglia, as well as other favourable districts, the problems of flax-growing and the moods of the plant became new topics for argument and dogma among the farmers. New machinery was evolved to help with the lifting. Many who thought flax altogether too temperamental to grow in peace time tackled and mastered it.

But among all these extraordinary demands for new crops, for bigger harvests, for better farming and improved methods of cultivation, the problems which arose created their own opportunities. They brought an excitement back to the land which infected everyone, even the lost men who had long since given up the struggle. These suddenly became aware of the intense activity of their neighbours. They saw that they were needed, that there was a point to farming after all; and some old spirit of competition or pride awoke in them again. One such old man, who lived alone on a poor 50-acre farm in the Peak district, tells his own story. In 1940 he was classed as a " C " farmer, which meant he was not getting the best out of his land and was therefore in danger of losing it.

" In 1914 this was as dainty a farm as you

could wish to see. I had three of my sons working with me then. We kept sheep and cattle and grew some tidy oats and barley. Then the boys went off to the war and didn't come back. The land got the master of me after that. I fell into debt, sold my cattle, and lived on rabbits and mushrooms and a bit of cabbage I worked myself. After a time even that got too much. I asked the moneylenders to help me with a tractor, but they didn't fancy it. The soil wore thin, blowing off the rocks, and bracken took a hold on everything.

"When the District Committee-man came in 1940 I was at the end of my rope. He said the land couldn't lie idle and he'd have to take over. I was mad. I almost fetched the gun to him. But we went for a walk round instead. My neighbour was ploughing down by the stone quarry and I said: 'Where did he get that tractor from?' The Committee-man said: 'From us'. I said: 'You just lend one to me—I'll show him a thing or two'. Well, I got one. I scrubbed up my old plough and cleared 30 acres of bracken that year—alone. I planted wheat, potatoes, and rye—alone. On the strength of what I'd done, the Committee advanced me enough money to buy my own tractor. I wheedled my widowed sister to come and live with me, and we got in a land girl from the village. Then we branched out. The Committee kept an eye on us and hired us more labour. Next year I cleared another ten acres of bracken. I gave it a good dose of fertiliser and planted a nurse crop which I fed off on young tegs (sheep). Now the land's clear; I'm growing rotations of wheat, potatoes, beet and peas. I've advanced my status as a farmer and I don't hear anybody saying 'poor old George' any more. I'm as young as any of 'em. I feel like a boy."

So profits the land by proper employment and change, giving us new crops, new ways, new farms—but also new men.

SUGAR RATION. Piled like coal in a factory yard, the raw beet awaits refining. The entire domestic sugar ration of the nation comes from this home-grown root. Cattle food and molasses are also by-products of it.

5

Reclaiming the bad lands

GOOD BRITISH farmland is a shrinking commodity. Consumed and lacerated by spreading cities and arterial roads, it has been shrinking for years. There was never so little of it as when the war began. And never before did we need so much.

The first thing war did was to seize a great deal more. Modern arms are insatiable in the matter of land; new weapons and new defences need land and more land, and generally of the best quality. Fighter and bomber aerodromes, requiring thousands of flat dry fields, obliterated many farms. This, of course, was unavoidable, but it did not end there. In wide rings round the cities, along the invasion coastlines, at strategic points everywhere, land was used up in essential preparations for defence and attack, in the siting of batteries, searchlights, camps, store-dumps, radiolocation, and battle-practice grounds. Since war began, several hundred thousand acres have been completely lost to food production in this way. Nor was that all; thousands upon thousands of good open meadows were mutilated by widespread anti-aircraft and anti-tank devices. Ditches were dug through wheat and broke through the underground drainage systems of many farms; poles, wires, logs, brick-heaps, concrete blocks and old cars were scattered over the fields in prodigal confusion. Necessary as these precautions were, they were at the same time a severe hindrance to cultivation.

If the farmer had thought fit to confine his attentions to the land that remained to him, ploughing up only so much as was contained by his fences, he would have reached a point of saturation when there were no fields left to plough. In that case we should have gone hungry, for the extent of British farmland proper, no matter how intensively cultivated, is inadequate to supply us with all we need.

But the farmer was not so content. He knew he must extend his territories and replace the lost lands by bringing back into use the inferior pastures which had fallen out of cultivation since the last war, and by reclaiming unwanted wastes of bog, fen and moorland which hitherto had resisted or did not repay cultivation. Such deserts lay scattered all around him. Many of these had never before been cultivated, because old-time methods could do nothing with them. They were the bad lands of Britain, and the farmer had long had his eye on them. Now the coming of the machine brought them within his grasp.

The reclamation of these idle acres was perhaps the most dramatic aspect of British wartime farming. Never before had this country seen such operations carried out on such a gigantic scale. It has been an achievement in tune with the scope of our

LAND OUT OF HAND. (1) Once a country lane surrounded by farmland—now a building estate. (2) Fields stifled by stagnant water through lack of proper drainage. (3) Dormant pasture, coarse with old grass; hill slopes covered with useless bracken.

time, conceived with imagination and carried out with the full-blooded use of the machine.

"After the fields, the swamps and commons, then the hillsides, then the mountaintops." That was the slogan. Everything outside the fields was a sort of desert warfare. And this campaign, waged with heavy modern equipment, may in its modest way be compared with the achievements of the North African army. In nine months that army subdued a coastline which once resisted the armies of Muhammad for a hundred years. Here in Britain the agricultural reclamation squads, assailing at times land so wild it had never known the plough, forced it to bear the first harvests in the history of man. It is here we saw the countryside's most remarkable transformation, and the real power of modern farming—in the harvesting of the mountains and marshes where no man had ever before seen a food crop grow.

It was to these extreme achievements, as well as to the humbler ones wrought upon common land and old pasture, that we owed so much of our security in terms of food. Potatoes flowering on the hills of Wales, oats and barley on Hackney Marshes, wheat from bogs, from scrubland, and once flooded valleys—these were but a few signs of the power of wartime farming. We saw all this; but, wherever there was open ground, we could see harvests taken; from ground long hallowed by tradition and custom—golf courses, race tracks, cricket pitches, bowling greens. These, covered with onions, cabbages, rye, potatoes, were not so much reclaimed as surrendered willingly to a necessary cause.

Britain, like the rest of Europe, was once a mighty forest roaming with wild pig, wolves, and deer. Our forefathers hunted these animals with bow and spear. Later they settled down, kept flocks, and began a primitive agriculture. It was they who cleared the forests and created pastures, for pastures are not natural to Europe. But the seeds of the forests remain. If the hand of man were removed altogether from the land, and his livestock wiped out, Britain and Europe would revert to their original forest state. The briar would run first through the rank, tough grass, bushes spring up and spread and tangle; for a time there would be dense jungle, then the trees would come.

At the outbreak of war it was possible, in certain districts, to see fields and patches of waste ground where this sinister process was already far advanced. One such place, a choked and notorious wilderness lying only five miles out of the city, was tackled by the Cambridge Committee in 1941. There were over a hundred acres of it, and it was very far gone. Neglected for 50 years, it was so overgrown with knotted vegetation that the place had almost the look of an African jungle. Thorn trees twisted their black crazy branches to a height of 12 feet. The sun filtered through but dimly. Foxes, rabbits, and badgers lived secluded lives there in continual twilight.

"This land wasn't worth more than twopence an acre," said a local man; "you couldn't shoot through it, you couldn't move in it, in fact nobody would go near the place. It was lost land all right."

To reclaim such a waste was a major operation, calling for plenty of labour and the best machines. Most of the machines were provided by our allies, and General Wavell supplied most of the labour. Late in 1941 Italian prisoners of war were set digging drains and cutting back the bush-tops. They worked well and seemed to enjoy it. One said he preferred it to being chased all over Libya, anyway. As the jungle receded before their axes, huge bonfires were built, blazing by day and extinguished by night.

The Italians were followed up by gyrotillers and bulldozers, tearing out the long tap-roots and dragging up the stumps. These, too, were gathered and added to the fires. The action of the gyrotillers threw the heavy clay into huge clotted lumps as if the place had been raked by intense gunfire. All this had to be worked and broken down. Land girls arrived with caterpillar tractors, rollers, disc-harrows and deep-furrow ploughs. They went over the ground again and again, battering and slicing those heavy clods, till they had reduced them to some semblance of a seed bed. The job was a wearing one; it took two hard winters to clear the land completely. But the result was a clean dry field, wide as an aerodrome, capable of producing hundreds of tons of wheat and potatoes—food for humans, not food for rabbits.

This job is typical. Similar large-scale efforts in all parts of the island have made farming history. Farmers will quote them for years to come and use them as a measure for the possible and the impossible. They will talk often of the Battles of Dolfor, Long Mountain, Stonyhurst, Hollow Moor, Swaffham, Feltwell and Burwell. For these are names which cover the country, from Cumberland to Devon, from the Welsh hills to the eastern Fenlands.

Take Feltwell, Swaffham and Burwell, for instance. They lie to the south of the Wash; they are part of the fertile Fen Country. Once they were names of desolation. They now speak for some of the most highly productive areas in Britain. Before the war, you would have to go far before you beheld a sight more drear than those wild marshes. Thousands of acres of low flat land smelling of rotten weeds and the sea. Nothing to be heard but the rattling of wind among the nine-foot reeds, the screaming of wild birds, and the slow sucking of the flood waters. A few pitiful signs of man's earlier attempts at cultivation merely added to the all-pervading sense of wilderness and decay. Broken dykes foundered among the morass, choked up and glutted with reeds and stagnant water. Tattered windmills, like huge dead crows, drooped motionless above the rushes. And here and there the black timbers of a deserted farmstead writhed in fantastic ruin as the spongy bog sucked at its foundations.

There had once been a hard-living community here, working the marsh as best they could. They had been driven out at last by failure and flood. High tides broke in and fouled much of the land. Reed, scrub and willow-herb took possession, and wild birds made a sanctuary of the place. Over the yellow swamp flew bittern, harriers, grebe, mallard, snipe, herons, redshank, and peregrine falcons. Coots and moorhens dived among the weeds. The marsh's only use to mankind lay in the tall rushes occasionally gathered for thatching purposes. It was a dismal enough prospect for reclamation; the area was the less fortunate heart of a particularly well-farmed district, and good black soil lay under all that waste. In 1941, however, in the face of great difficulty, 1,500 acres of Feltwell Fen were successfully reclaimed by the Norfolk W.A.E.C., and first-class harvests were subsequently taken from it.

It was about the same time that the Cambridgeshire Committee approached Swaffham and Burwell. Their task was, if anything, even more forbidding. They knew it would call for everything they had got. Before they could bring their machines to the land, concrete roads had to be built—pushed out across the bog like duckboards. They had to be built on very solid foundations, and the remains of many a bombed East Anglian cottage went into these.

The roads were laid out in large squares, isolating a hundred-acre block at a time. From these the machines worked inwards, beginning with the drains. For every hundred acres of land, five miles of drains had to be cut. The existing system, a complex net of high- and low-level dykes, had to be cleaned, strengthened and repaired. This, before any other work could be done. Slowly the land began to dry out. The gangs waited with the necessary patience.

If the tractors had started work too soon they would have been hopelessly bogged, or perhaps swallowed up altogether. But at last the place was firm enough and the onslaught began. A large force of workers, among them 60 land girls, set about the reeds with knives, cleared them and burned them.

Then came the caterpillar tractors drawing the heavy assault machines—Canadian prairie-busters, disc-harrows, and the formidable Australian five-furrow stump-jumpers. The blades of the stump-jumper are designed to spring clear of submerged obstacles, such as rocks and logs, instead of breaking off as the blade of a normal plough will do. It proved, on this particular job, a very useful tool to have. For it discovered quite soon

BREAKING NEW GROUND. Four stages in reclamation: A 22-ton military bulldozer, *top left*, launches the offensive, slicing down trees with its razor-edged nose. *Top right*, a gyrotiller works through the smaller bushes, dragging them out by the roots. *Middle*, huge clods of wild ground being thrown up by a stump-jumper plough. *Bottom*, the clods are broken up by heavy discs ready for sowing.

that parts of the Fen were packed with submerged obstacles, huge trunks of petrified bog-oak, unyielding and hard as iron, lying around like outsized coffins in a churchyard. Buried only a few inches deep, preserved by peat and salty water, they were relics of a primitive forest and had lain there for thousands of years. No one had ever tried to shift them before. Here and there some old farmer had stumbled upon one with his plough, had hacked at it ineffectually with an axe, but that was all. Now they were coming up; their graves were needed; but clearing them was by no means an easy task. Said the gang-foreman in charge of the job:

"They nearly broke our hearts at first. Some were 7 feet thick and as much as 100 feet long, and they were lying 50 and 60 to the acre. We couldn't plough an inch till they were shifted. We decided the only thing to do was to dig round them and then blow them up with dynamite. A hundred land girls—Lancashire lasses mostly—set to with spades. They worked like Trojans, but they had to dig deep and the job took months. After they'd laid them all bare the Royal Engineers came, drilled holes in the trunks, set charges, and blasted 'em up into short lengths. These the girls towed out with tractors and piled along the roadside."

There they remained, torn from their ancient beds, vast black hulks crumbling with dust and age. They looked like those charred wrecks of German aircraft which were dumped in the Kentish fields after the Battle of Britain. They were in fact a symbol. For, ever since man first cultivated the land, they had strangled this part of the Fen. Now the fine black soil was rid of them at last, and the highest yields of wheat, potatoes and sugar-beet could be taken from it every year.

That is only a part of the marshland's story. But let us turn now to the hills, where an altogether different problem was tackled. The mountains of Wales, the moors of Devon and Cornwall, the Pennines, and the hills of the Lake District cover nearly a third of England and Wales. In the past

BREAD FROM THE WILDERNESS. Feltwell Fen, was a dead place of broken dykes, sour water, and yellow reeds. Now the dykes have been cleared, the reeds replaced by acres of lusty wheat.

DIGGING UP THE PAST. Oak trunks, five thousand years old, were found buried in Feltwell Fen. They were dug out and dynamited to free the land for the plough.

they had provided little more than grazing for sheep and the hardier breeds of cattle. The difficulties of ploughing this high land were, of course, great, for much of it has no more than a very thin coating of soil over the rock. But on some of the rounded hills of Cumberland and Wales there is a body of soil that goes deep and carries an abundance of bracken. And there is an old tag which says, " Where bracken will grow, potatoes will grow ".

The Montgomeryshire W.A.E.C. had this in mind when they reviewed the bleak heights of Long Mountain and Dolfor. Nothing lived up there but a few sheep and a few wild ponies. There was a bit of dead grass and a lot of scrub. But the Committee thought that, given the proper equipment, they could put crops on those hills; there was no precedent for such a venture, but the time was spring 1940—a good time to ignore precedent. So they made their plans and the " Montgomeryshire Experiment " was begun. It was the first large-scale reclamation of the war.

The Committee foreman-in-charge, a tough, jaunty little Welshman with a jaw like a spade, had the job of organising the actual operations. For three years he spent the better part of his daylight hours upon those mountains. He said :

" When first we started, there was no model for what we intended to do. Whether the hills could feed anything save hawks and wild ponies was something most people doubted. But those were Dunkirk days, so we thought we'd try it out. We started in on the side of Buttington Hill—part of Long Mountain. Bracken grew ten feet high, higher than the tractors, and thorn bushes

UPHILL WORK. The first stage of the Montgomeryshire Experiment. A prairie-buster climbs Long Mountain, turning up great furrows matted with bracken roots. In the distance stands one of the "Gates of Wales".

had to be pulled out by steel hawsers. The hill was steep; we ploughed it, disced it, and planted potatoes right away. That was our first mistake—we should have taken a pioneer crop off first, but we wanted potatoes and we were too impatient."

There was little labour and fewer machines that year. The potatoes were planted and lifted by schoolboys and village women. Next year the bracken sprang up as thick as ever. But this time the Committee knew how to deal with it. New machines had arrived, prairie-busters and heavy discs. With these they tackled the bracken in time and ploughed deep.

In 1941 they took in more of Long Mountain, ploughing 400 acres right along the peak, farming now well above the clouds. Here they made sure of the bracken. The sod was turned over flat and given several discings. They took tests for manurial deficiency, and gave the land a good dusting with lime and basic-slag. Next, instead of immediately sowing potatoes, they sowed a crop of rape and fed sheep off it. The sharp hooves of the feeding sheep, together with their droppings, helped further to build up the soil's fertility. By the following spring Long Mountain was ready for potatoes.

From Long Mountain the Committee moved on to Dolfor, high land on the edge of Radnorshire, and took on a thousand acres there. After that, they extended their hill territory by nearly a thousand acres a year. They were farming on a scale unknown to the more confined farms of the lowlands. They were cropping 500-acre blocks of land at a time. There are " fields " such as these, running across Dolfor, where the ploughs were cutting single furrows 1½ miles long. They got the measure of the job. They covered the mountains with crops, and regrassed the highest, most desolate peaks, with all the confidence of a farmer working his favourite meadow. And from the potato-clamps they set up—storage mounds of earth and straw which stand on the hills throughout the winter—sufficient potatoes were taken to feed the whole of Manchester.

The Montgomeryshire Experiment proved without doubt the possibilities of those once-neglected hills. It proved that, with the proper use of machinery and fertilisers, good crops can be grown well above the traditional contour limit—to heights of a thousand feet and over. It proved that even higher levels can be resown with good grass that will thrive and maintain stock on land which was hitherto quite unproductive. Such achievements, apart from their extreme usefulness in war time, are likely to have a profound effect on the future of hill-farming.

The efforts of those pioneers were echoed in many other parts of Britain too. Ten thousand acres of rough Pennine moorland were ploughed for kale and oats, and fodder for cattle. In the North Riding of Yorkshire, ploughs and cultivators worked in cloud and snow while the valleys beneath them were bathed in sunshine. Near Ullswater, in Cumberland, tractors and threshing tackle were edged up the 1 in 4 gradient of Hallin Fell to harvest the highland crops and confound the neighbourhood, which never believed they could get there. On the chalky switchback Sussex Downs large areas, cleared of gorse and bracken, carried the first crops since the days of the Saxons. In Wiltshire, 500 acres of Kings Heath Common, whose grazing rights have been handed down from family to family since the days of King Athelstan, were ploughed up after a council of war held by the Commoners in the ancient Court House of Malmesbury. At Northwood, Slindon, hundreds of acres of bramble, willow-herb and scrub, riddled with rabbits and foxes, were cleared up and the vermin slaughtered by gas. John Bunyan's " Slough of Despond "—a low, sterile swamp lying between Ampthill and Bedford—was drained and cleared and 2,000 acres of it sown with wheat and potatoes.

These are *new* lands ; and they are only a fraction of what was reclaimed in the war years. They replaced the building sites, the airfields, the camps, the necessary wastage of war. The food they gave us was our absolute gain. That food was here, it could not be torpedoed, and it cost no lives. What it did cost was imagination, sweat, skill, and the long, back-breaking, dawn-to-dusk, all-weather endurance of the farm worker—man and girl.

THE HILLTOP YIELDS a 2,500-ton harvest of potatoes, fruits of the first large-scale reclamation of the war, which turned Long Mountain from a place of bracken to productive farmland.

Responsibility for most of the large-scale reclamation fell, of course, on the County Committees, for they were a national organisation with special facilities for such work. They could command equipment, capital, and expert knowledge outside the scope of the private farmer. But much, all the same, was achieved by him too, and the more remarkable it was because of his limitations. Here is a final story, told by a Hampshire man of what he did with 1,000 acres of scrub in that county:

"When I took on this land," he said, "I reckoned it the most worthless stuff on God's earth. It was Downs-type land, steep and rough, and covered with bush as black as a monkey. One of our heifers went and died in it and we didn't find her for three days. The last chap here used to live on blackberries and rabbits: I don't know how he did it. When I arrived, they said to me, 'That land won't grow anything, mister. Why, it's so poor you got to put straw out in the winter to keep even the rabbits alive'. I started anyway. I got some stiff tackle together and we tore out the bushes and burnt them. I set my men killing rabbits at a tanner a time and they cleared over 2,000. Then we ploughed it and gave it a good dose of lime. I re-seeded a hundred acres for dairy cattle, and put sheep on to the rest. Next year I planted wheat. Since then we've had oats, barley, rye, sugar-beet, and spuds off it. It's pulling its weight now, but it was mighty poor stuff to begin with. Why, before the war I wouldn't have looked at it—not even at no rent. Only Hitler and the War Ag. drove me to it."

THREE-MILES HARVEST. In 1939 Hollow Moor was derelict land, feeding a few cattle and a few wild ponies. In 1943 it looked like this, a vista of golden stooks, bearing 350 tons of oats.

6

The countryman's myriad foes

THE SOIL, by its nature, is not always benevolent. It can bring forth in equal abundance both good and evil; it can produce a good crop and at the same time the means to destroy it. Earth-born pests and plagues are the enemies within the gates—microscopic but formidable saboteurs whose existence depends wholly upon the damage they can do. They are a potential menace in the best soil and the farmer can never let up in his struggle against them. His crops and his cattle are in continual danger from them. They attack from all quarters: from the air, the ground, and under the ground. They take forms both visible and invisible, from the rat and the woodpigeon to the blight-fly and disease-producing virus. In war time, every resource of farmer and scientist had to be mobilised against them.

Certain wild birds, for instance, are a serious threat both to the newly sown seed and to the harvest crop. They are elusive, and well equipped by Nature for sudden attack and quick get-away. Woodpigeons, particularly, are insatiable marauders and very cunning. When planning a raid they carry out very careful reconnaissance beforehand, and are at all times extremely sensitive to danger. But when they get among the crops they eat with astonishing rapidity and do great damage. The gizzard of one dead bird, shot down in a Home Counties field, was found to contain over a thousand corn seeds, black oats and buckwheat seeds. And he was no exceptional glutton.

To counteract the ravage of woodpigeons, large-scale campaigns were organised against them. Army observers and members of the Royal Observer Corps, switching their attentions from enemy aircraft, reported their movements, and huge flocks were ambushed and shot up from such information received.

A notable expedition was organised one year in Oxfordshire. Large flocks had been reported roosting in a certain wood. Farmers, Home Guards, soldiers, anyone who could fire a gun, went out and lay in wait for them. Dead birds and wooden dummies were used as decoys. When the birds arrived, weaving in the air, the sight of the decoys lulled their suspicions. They came in to roost with absolute faith, the guns opened up, and there was terrific slaughter.

House-sparrows are capable of causing great havoc among standing wheat. When the corn is ripe, they move out from the towns in numbers and, joining the local sparrows, they will come down in a scattered cloud, spread out, and in a short time leave thousands of empty wheat-ears transparent against the sun. At times the tough, belligerent house-sparrow is a menace to that useful bird the house-martin, which feeds upon insects. Many house-martins' nests are commandeered by thieving sparrows in spring and summer. Shotguns and airguns had little effect on the sparrow. It fell largely to the shouts and rattles of schoolboys to keep him at bay. This the boys did with enthusiasm, spending long hoarse hours in the fields at harvest time.

Then there are rabbits. Their combined operations throughout the year costs the country many millions of pounds. The flesh value of the wild rabbit in no way compensates for the food he eats. He has, therefore, been the object of a sustained and well-organised punitive expedition, embracing methods of extermination unknown to his ancestors. In the early part of the war he was in sole possession of much valuable land. Before long he was being trapped and gassed in his millions.

The chemical gassing of rabbits is comparatively a new thing; at least, it had never been practised on such a scale in this country before. It is invaluable for attacking rabbits underground. It is ruthless, quick, and effective. One farmer, tired of taking pot-shots at a warren multiplying in one of his fields, appealed to the local Committee for help.

"They sent me along a tin of powder," he said, "with instructions. It was queer-looking stuff; I didn't see what good it would do against all those rabbits. Thousands of them there were, they'd eaten me right down to the doorstep. Well, I gave it a trial. I took my boy up to the bank and we fixed the gas and stopped up all the holes. I don't know what happened, but we never saw them again."

The rural rat, who spends his summer in the hedgerows, banks and ditches, and moves when the time is ripe into the stocked barns and wheat-stacks, is another pest capable of enormous destruction. One pair of rats can produce 880 surviving offspring a year, and each will eat 10s. worth of food. Such a menace are they that the Government has had to enforce, by special orders, the taking

ROGUES' GALLERY. The squirrel, rat, woodpigeon and rabbit attack from trees and sky and from under the ground. They will eat any crop from the field or the barn. They cost the farmer millions of pounds a year.

of precautions against them. At threshing time all ricks must be fenced around with wire to prevent their escaping. There may be a hundred or more rats living in one rick. When the corn is threshed, they are forced into the open and can then be polished off with sticks and dogs.

The Pest Control Staffs of the County Committees trained special operatives whose full-time job it was to fight rats all the year round. The familiar rat-catcher, with his stick, gaiters, struggling bag of ferrets, and quick wiry terrier, was still an important public servant, but he had now been joined in his work by a figure much less familiar, though equally dangerous to rats—the trained land girl. The isolated, intermittent attack upon the rat is usually a waste of time; it merely drives him away to fresh ground. The assault must be a planned, co-ordinated drive covering a wide area. Many such operations were organised by the County Committees, through their Pests Officers, and entire colonies of rats, numbering many thousands, were wiped out in this way.

The land girl rat-catcher (and there were many of them) must be one of the most intriguing personalities the Land Army produced. She took to her job with a determination and lack of fastidiousness which must have astonished many of us with old-fashioned ideas as to woman's antipathy to rodents. Some, working entirely on their own, established remarkable personal records. In two days, a 19-year-old ex-dress-designer from Leicester gathered 327 carcases from a Yorkshire granary: and 300 rats can eat three tons of wheat a year.

Here is a story, told by an East Anglian farmer, which illustrates something of the impact of this new type of rat-catcher on agricultural life:

"A girl came into my yard one morning and said to me, 'You've got rats here, haven't you, mister?' and I said, 'Well, maybe half a dozen, but they won't hurt you,

RATS BREED, LIKE DISEASES, in the midst of what they feed upon. From the rick, *above*, a trained land girl hunts them out with ferrets. *Below*, the result of one day's offensive in a Hereford barn.

missy", and she said, 'It's not me I'm worrying about'. She took a look at the barn and the drains and said, 'You've got more than half a dozen, mister, but leave 'em to me, I'll fix 'em'. Then she opened a sack and started bringing out bait and poison. I said, 'What d'you think you're up to?' and she said, 'Don't you worry, mister. I'm the rat-catcher. The Committee sent me down'. But I did worry. I flew around to my man and I said, 'Lock up all the stock, Alf, there's a female rat-catcher here and she'll slaughter the lot!' He was as frightened as I was; but we needn't have bothered, the girl knew her job all right. Later that week she came to me and said, 'Take a look at the barn, mister'. I did, and there were 60 dead rats laid out in the prettiest pattern you ever saw. And I never knew we'd got that many in the whole village."

The grey squirrel, charming as he may appear to the eye, is but another of the nation's food robbers and must be ruthlessly destroyed. Since establishing himself in this country he has done immense damage to crops of many kinds, and to the farmer he is nothing more than a tree-rat and a dangerous and destructive animal.

Other less obvious pests, diseases and blight, attacking both plant and animal, are not so easy to fight. They are intangible and work invisibly. Wireworm and leather-jackets, eating away the roots and rotting potatoes; the whitehead fungus which rots the cornstalk; the bulb-fly maggot which breeds in the young shoot and withers it; frit-fly and turnip-fly whose parasitical presence can turn a lusty crop into a heap of infected rubbish; these are only a few of the plagues which the land, like the human body, can contract without warning. Then there are those deadly, extremely contagious diseases to which even the healthiest of cattle are subject; diseases which can appear in a strong herd overnight, carried by birds or rats or even by a piece of infected straw; foot-and-mouth disease, mastitis, sterility, contagious abortion, foot-rot in sheep, "sway-back" in lambs.

ENEMY ACTION. Left, oat-sacks ravaged by rats. Right, a field of collapsed wheat infected by the fungus disease "eyespot".

CHEMICAL WARFARE. Spraying fruit trees against insect damage. The men wear protective rubber. One, perched in a crow's-nest, plays on the upper branches out of reach from the ground.

All these are natural catastrophes against which the farmer must always be prepared. For unless he knows how to guard against them, or treat them, he may lose an entire harvest grown from irreplaceable seed, or a pedigree herd which may have taken him years of careful breeding to build up. The scientist has given the farmer great help in this battle and his efforts in discovering counter-measures to combat the threat of pest and plant-sickness have been as exciting, in their own way, as the measures taken to beat the magnetic mine. Intensive research into the habits of certain pests has provided the farmer with completely new weapons ;

has established, too, new ways of maintaining the health of the soil so that its resistance to disease should be increased to a maximum.

Improved methods of soil testing, for instance, reveal just what fertilisers are lacking in its make-up. The farmer, hitherto largely dependent on trial and error, now knows exactly how much lime and fertiliser the land needs to maintain its balance of health. Insecticides have been standardised. The chemical constituents of the derris insecticide, for example, have been isolated and identified, and its wasteful properties eliminated, so that the remaining product should be specially effective.

The sinister wireworm is difficult to attack directly. But methods of estimating its population in any given field have been discovered, so that the farmer may know beforehand what crops he can sow with safety. If the wireworm content is high, he will sow only those crops that are immune, such as beans and flax ; if medium, he can grow barley ; if low, and only then, will his wheat and potatoes be safe. By this pre-knowledge the farmer is able to control the extent of damage the pest will do, and so avoid any repetition of those last-war disasters when ravage from wireworm became, at times, so serious a menace.

Many plant diseases are being eliminated from the outset by seed-dressing and inoculation. The turnip-fly (which attacks kale and cabbage, too) is discouraged by a mixture of kerosene, naphthalene and benzine. Cereals are rendered immune to the withering Smut and Stripe diseases by dusting the seed with an organo-mercury dressing. All these precautions are just a fraction of the hundred-and-one essential bits of information the ordinary farmer must have at his finger-tips. It is all part of his heavy load of care.

As for the health of fodder-crops such as lucerne, clovers, peas, beans and vetches, whose failure to flourish in certain districts was often such a mystery to farmers, scientists have discovered that the bacteria necessary for proper growth were often completely lacking in some soils. By inoculating the seed itself with the appropriate bacteria this deficiency has been met, and the farmers are now getting much of their seed already so treated.

Lime is essential to grass and fodder, and for neutralising acid soils. There is much natural lime in England; but in cultivated fields it has to be continually replaced, for rain washes it away, and animals take it for bone and do not return it. The farmer relies for most of his supplies on burnt lime, ground limestone, marl and chalk from the quarries, and on waste lime gathered from paper and sugar-beet mills. To help in the increase of fertility a Government scheme had been in operation since 1937 by which farmers could obtain one-half the cost of lime or one-quarter the cost of basic-slag purchased and used to improve their land. This scheme had far-reaching effects.

A better knowledge of the value of lime and of artificial fertilisers to supplement dung saved the farmlands of Britain from serious exhaustion in the war. Supplies had to be weighed carefully against the farmer's need, for there was no surplus. And he, patiently and with imagination, had to supplement his ration by turning as much of his straw as possible into farmyard manure. Only by his resourcefulness, and by the economy of the scientist, could the land have been kept healthy and in good heart through five years of shortage and overwork.

TO KEEP THE LAND SWEET, lime, which is consumed by growing crops, must be continually replaced.

7
Green pastures

MILK IS Priority Food No. 1. It is the most complete food Nature can give, and one which our land and climate alike are well able to produce. In view of this, we turned naturally to milk to replace many other foods we could not get, to compensate for the lack of eggs, butter, meat and fruit, and to maintain the health of essential workers and fighters. The cry for milk was insistent from the very beginning of war—milk to check malnutrition, milk for the mines and factories, milk for children, for invalids and nursing mothers. We were, in fact, consuming more fresh milk than ever before in our history. And in spite of the lack of imported animal feeding-stuffs and the restriction of grassland by ploughing, the farmer was selling more than he had ever sold.

How was this accomplished? It was due to the great switch-over on the part of many farmers from other forms of stock-keeping to dairy farming; the growing of increased acreages of fodder crops; improvements in dairy practice, herd conditions and methods of milking: but, as much as anything else, it was due to a revolution in the use of grass. This revolution became possible largely through the widespread adoption of the "ley" system of farming, a method long advocated by Sir George Stapledon, one of our foremost agricultural scientists.

Ley farming is a system of rotational cultivations aimed at restoring fertility to exhausted fields, and rejuvenating old pastures, by sowing them with temporary grasses. At the same time it is a method of providing young healthy grass for the feeding of cattle. It would be difficult to over-estimate the profound effect its adoption has had upon British agriculture, or the effect it is likely to have upon its future. It was not a new theory. It had been practised for years in certain districts of Scotland, Devon, Cornwall and elsewhere, where climate and conditions are specially suitable. But to farmers in general it was an innovation, a system with unusual possibilities, and it gave them a new respect for grass.

Grass is a crop, and can be sown and harvested like any other. Moreover, it gives back to the land qualities which other crops take from it. Fields grow tired of wheat; they are sapped and destroyed by it; if sown with nothing else, their yields drop away. Farmers have been aware of this since medieval times. In the old days they rested fields by letting them lie fallow for a year or two. To-day, however, we cannot afford the idleness of fallow land; the resting and restoration of fields must at the same time be productive. Grassing may prove the answer to that problem.

But what exactly will it mean to the farmer? It may mean the end of permanent pasture altogether. For old grass is wasteful and often dangerous. If neglected year after year, it deteriorates in quality; tough weeds oust the more nutritious plants, insects multiply in the ground, and disease-producing parasites cling to the undisturbed turf, threatening every animal that comes upon it.

The ley system is more than a purifier of land; it is bringing a completely new rhythm to the farm. Grass becomes mobile; it goes from field to field, following the plough and taking its turn with the other crops. As the arable fields show signs of exhaustion, they are sown with a one- to five-year ley of grass. The cattle are moved on to this from the once permanent pasture, which in its turn is ploughed up and sown with crops. In this way the whole productivity of the farm can be increased, the health of the soil is toned up and the cattle get better fodder.

It sounds simple enough, and the theory in fact is simple, but successful regrassing depends upon carefully balanced seeds-mixtures, as well as skilful management of the subsequent crop. Grass is a broad term embracing hundreds of plant varieties, all with their own qualities and characteristics. You cannot sow any grass just anywhere for any purpose. Your special need demands its special formula. And it is here that the scientist steps in again. Research students at the Welsh Plant Breeding Station and elsewhere have for many years been breeding special strains of grasses for all purposes, climates and localities. They have evolved grass to suit highland and lowland, grasses for wet and dry areas, precocious young grasses that leap up in a few weeks and give winter-weary cattle a green bite in early spring, and other types that take longer to mature. They have selected wild grasses, crossed them, bred them, and produced pedigree strains unknown to this country before. They have imported valuable types from the pastures of Australia, New Zealand and South America, and have even recovered

FROM COUCH TO CLOVER. Left is a picture of old pasture, coarse, dry, and choked with weeds. On the right it has been ploughed and re-seeded with a mixture of new grass and clovers, trebling its food value.

GRASS IS A CROP.   Cut in its prime, knee high and red with clover, then dried in the summer sun, English grass, when good, is the best in the world.

and reintroduced to Britain certain native species which were taken abroad by emigrants over a hundred years ago and have since developed completely new characteristics. These grasses are able to play a valuable part in supplementing the more standard types that have been mixed and sown by the British farmer in the past.

Ley farming offers many advantages. It makes it possible for the cattle to go to the regrassed pastures much earlier in the year, and stay on them much later than they could upon old grass. During the lean winter months they can feed upon mixed fodders which have been specially sown, cut and preserved by means of ensilage. Ensilage is a method of storing young green fodder so that it retains most of its original goodness.

The silo, once a rare sight in this country, can now be seen on many farms. Often it is built of moulded concrete, drum-shaped, like a diminutive gas-holder. But it takes many other forms too; it may be of wire and stiff cardboard, or perhaps an old iron tank—one Lincolnshire farmer built his of railway sleepers lined with newspaper. The art of making ensilage is not easy, and it took the war to spread the practice. The fodder must be cut at exactly the right stage of growth and fed into the container just so much at a time. It must be watched continually to prevent evaporation and over-heating, and the addition of molasses must be adjusted carefully to its condition. When you saw land girls dancing round the top of a silo, they were not merry-making but treading the grass to compress it; this is a process rather like treading grapes and must be carried out with equal care.

But, complicated as it all sounds, many farmers speedily developed an instinctive silage sense; they learnt to judge its quality by colour and scent alone, and evolved their personal variations in the making of it. Though County Committee experts were eager at all times to give advice on re-seeding and silage-making, there were many farmers who took a special pride in compounding their own mixtures, discussing them at length with their neighbours, making claims, boasting of secret proportions.

In one way and another, therefore, by

TREADING THE SILO. One stage in the making of ensilage, a method of preserving green fodder.

Grass, clovers, lucerne, peas, beans and vetches, when cut, are placed in a large airtight container—the silo. Here they are compressed and watered with molasses to prevent fermentation. The finished product is much better than hay; it is rich, oily, and pungently aromatic, looking rather like gold-cut tobacco, and providing a food with all the concentrated nutriment of cow-cake.

embracing those new ideas, by growing roots, cabbages and kale in greater abundance, and by utilising every scrap of green-stuff about the farm, the farmer was able to bridge the bare months from September to March and keep his herds fed. And throughout this complex struggle he was aiming at high milk yields both winter and summer.

But vital as it is, balanced feeding is only one element in milk production. Breed is even more important. You may feed an inferior cow on the fat of the land, but her return in milk will be no better than her nature. Owing to indiscriminate breeding there were many such animals in this country at the beginning of war—beasts of inferior ancestry, with bad traits and big appetites, who were not worth the land they took up. A cow is not a machine; one will eat as much as another, but the amount of milk given by each varies widely. Every effort, therefore, had to be made to improve stock quality throughout the dairy industry, to cut out such passengers as there might be, and to concentrate on the higher-yielding animals. This was done. In place of the occasional mixed herds of happy-go-lucky, lump-backed, ill-conditioned animals, you could now see more and more clean, beautifully formed cattle coming on to the land; tawny, heavy-uddered Jerseys, red and roan Shorthorns, strong-limbed Ayrshires, and smooth-flanked Friesians dappled with black and white.

Heifers are judged from one standpoint only — calf-bearing and milk-production. Their looks, build, stance, shape of hind-quarters—all point to this. But only after calving can their value be properly assessed. Milk-recording, a regular survey of the yield and butter-fat content of each cow's milk, enables the farmer to isolate the valuable animal from the passenger and to adapt his breeding policy accordingly.

The aim and ambition of every farmer are to establish his own pedigree herd, bred to his own requirements. The days of indis-

SUMMER GRASS for winter feeding can also be preserved by an artificial drying process. Cows must be fed well or their milk yields will decline.

THE FARMYARD DROWSES in the midsummer afternoon. Slow, heavy with grass, red and roan Shorthorns amble in from their pastures ready to be milked.

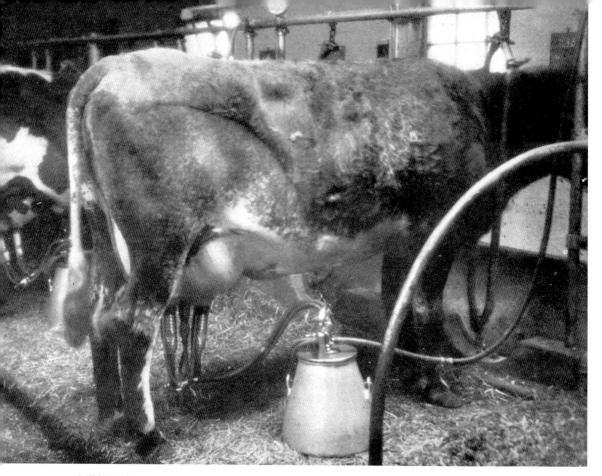

MECHANISED MILKER. Clean, swift, time-saving, this machine milks by compressed air. And cows, being creatures of temperament, like it; its gentle rhythm soothes them.

criminate bulling are numbered. Bulls are now carefully judged on the qualities of their progeny. A bad bull goes swiftly to the slaughterhouse, but a good one is prized above all else, for on him the herd is built and from him comes all its distinction.

But, because so much depends on the right bull, they are expensive and not everyone can own one. In some districts farmers clubbed together for a first-class animal and shared it out on a co-operative basis. Artificial insemination, developed so successfully in Soviet Russia, was tried out, too, on an increasing scale. In this way a good bull can be used to the best advantage, and can inseminate distant herds, by means of the test-tube, at the rate of over ten cows a day.

A few stations were set up to experiment with this method, which is likely to have a profound influence on the future of stock-breeding in Britain. In short, the bull, and the way he is employed, have become the responsibility of the whole farming community, and bad bulls are no longer tolerated as they once were. For there is no worse tragedy, when you are building up a good herd, than for the wrong bull to be found running with the heifers.

That the children of this country were spared the horrors of malnutrition is due very largely to the extraordinary energy of the British dairy farmer, and to the Minister of Food, who guaranteed that all children under the age of five and all expectant mothers should get their pint of milk a day. He would not, he said, see the future generation suffer for the follies of the present. The farmer seemed equally determined.

In the 1914-18 war, milk-production fell by some 40 per cent. There seemed no reason to suppose it would not suffer equally in this one. Yet, in spite of severe conditions and hard winters, the consumption of liquid milk broke the highest of peacetime records. Some people may be surprised at this; they were almost certainly not getting quite as much milk as they would have liked and they may wonder who was getting it. The children and the mothers were getting it. The rest of us could survive a shortage; these could not, without danger. Nevertheless, the consumption of fresh milk in 1944 showed a 40 per cent. increase above that of pre-war.

Winter milk-production was one of the farmer's greatest problems: summer milk can usually take care of itself, but winter is always a time of shortage. Many steps were taken to reduce this margin, chiefly by the feeding of specially grown fodder and silage already mentioned, and by arranging for late-autumn calving so that the cow might meet the winter with a natural flush of milk.

Dairy farming is no fun at the best of times; you work through all the hours of daylight, seven days a week. Cattle need care and attention whatever day it is. In winter you are up long before sunrise. At calving time you are often up all night. You milk twice a day, dawn and afternoon. You have to wash down each animal, feed it, milk it, clean the shed, clean yourself, and ensure the purity of the milk, over and over again, day after day. If that were all, it would be quite enough. But many dairy farmers, unused to cultivation, had to take that up too—planting roots, growing fodder, making silage, in order to provide their cattle with food that was once dropped at the door in bags. They did not assume these new responsibilities just for the love of the thing. Or for the money. Their work was real war service.

FILTERING gives clean milk. Recording each day's yield distinguishes the good cow from the bad, and shapes the farmer's breeding policy.

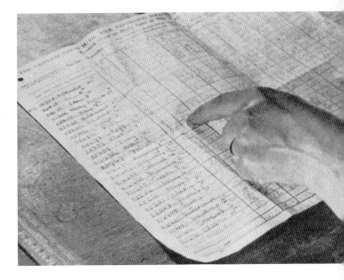

Meanwhile, with the better class of beast beginning to dominate the herd, with standards of health, cleanliness, feeding and methods of milking all undergoing revolution, there was being built up in this country a dairy industry that had never been in better shape. The farmer had behind him a vast national organisation, the Milk Marketing Board, his own creation, which regulates in peace time farmers' sales of milk and the prices they receive, and guarantees a sure market. The Ministry of Agriculture for its part had set up a national health service for cattle, while officials of the W.A.E.C.s were always ready with reliable advice and assistance on matters affecting milk-production and cow-keeping generally.

The high stage of development reached by the dairy industry in this country was not achieved by letting something else go hang. On the contrary, it was planned and adapted to suit the whole economy of the farm, and had a definite influence for good throughout all its branches. Re-seeding of grassland, the increase of home-grown fodder, the building up of first-class herds with regular health inspections, all these activities worked in with one another to attain a higher standard of production all round. The extraordinary wartime increase in milk-production was but one sign of the vitality of present-day agriculture.

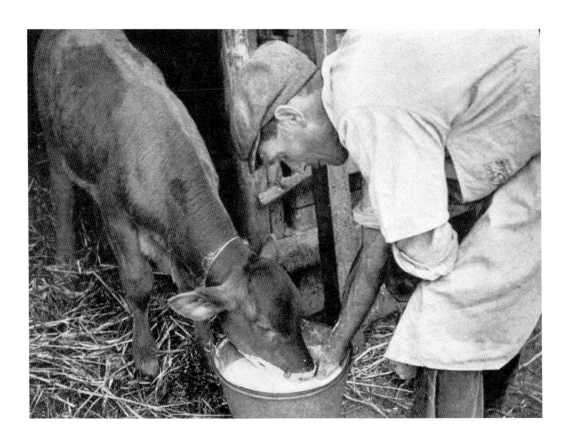

8

Farming in Scotland

SCOTLAND is not a rich country by nature. In area it is small, about one-sixth the size of Britain, yet it contains many types of landscape and a diverse range of soils. From the rolling green Cheviots and the swift tides of the Solway Firth away north to the shaggy Shetlands in the latitude of Labrador, and from the fertile fields of its eastern coast westward to the rain-swept, wind-blown Hebrides, no comparable area in the world perhaps can show a greater variety of agricultural conditions. All types are worked there: potatoes and grain grown in record crops on the old red sandstone of East Lothian; sheep- and cattle-raising on the southern uplands and the highland hills; and the meagre and often still primitive cultivation of tiny crofts scattered over the peat moors and mountain valleys of the far west, where Atlantic gales beat relentlessly on a struggling vegetation.

In many parts of the country agriculture is no easy business. The natural beauty of Scotland is prodigious and laid out with a lavish hand, but in soils that are kindly and easy to cultivate she has not been so generously endowed. Neither has history made the task of her husbandmen any less exacting. Sir Walter Scott, in his *Tales of a Grandfather*, says that " The generation of which I am an individual . . . have been the first Scotsmen who appear likely to quit the stage of life without witnessing either foreign or domestic war within their country ". And fighting and farming go ill together. But as the days of internal strife and border warfare receded into the past, Scotsmen turned themselves to the conquest of the soil and won many victories, often against heavy odds. Land was laboriously reclaimed and patiently tilled, new methods were adopted, livestock was improved. Within the limitations of its cultivable area, Scotland achieved high levels of farming and food-production.

But that area is limited indeed; of Scotland's 19 million acres only about $4\frac{1}{2}$ million are tillable; of the rest, $10\frac{1}{2}$ million acres are rough hill grazings, while a good part of the remainder is mountain land capable of supporting little but deer and game.

Consequently, war found few Scots farmers in a position to make spectacular gains in output. They did not have great stretches of rich permanent pasture, with its accumulated fertility waiting only for the plough to transform it into human food. But they were able to break up such old grassland as they had; they could reduce, too, the proportion of grass on their arable land; and they could intensify cultivation all round. These things they set themselves to do, immediately and with energy. Their plans had been prepared in good time.

On the first day of war the Scottish

THE BANKS OF THE TWEED. Pine woods and walled fields, heather and grass, sheep on the slopes and salmon in the river. Such is the eastern border country of Scotland.

Department of Agriculture was able to call together some of the best farmers of every district to form the local Agricultural Executive Committees. These Committees, armed with their special powers, operated similarly to those in England, save that they were even more decentralised. Their first job was to increase cultivations wherever this was possible, but their plans had to be adapted to suit different districts, for Scotland is ill-balanced and intensely regional in its geographical lay-out, and no one plan could cover the country as a whole.

There are three main branches of farming in Scotland, dependent upon the distinct character of three areas. In the east, from the Moray Firth down to the Berwick border, lie the fertile lowlands of the coastal

sheep and cattle on the heights and tilled by crofting communities in the lower valleys. Each of these areas lives a special life and has its own local problems, so that it had to approach the war in its own way. Each was able to adapt itself and make its special contributions.

The green arable belt in the east has always been a natural cropping country. The land slopes up from the sea westward to the hills and remains good, much of it very good indeed, until the higher levels are reached. Here the rearing and fattening of livestock have long been complementary to the growing of crops. Yet by the outbreak of war, in spite of the fertility of this soil, only one-third of it was cultivated. Hundreds of thousands of acres had gone back to grass and were being dominated by more and more cattle and sheep. With much of the rest of Scotland rough and resistant to the plough, the task of supplying the necessary increases in cultivations, which in England could be taken from almost anywhere, fell largely to this area alone.

Expert calculations prove that whereas 100 acres in beef or mutton can feed nine people for a year, the same area in milk can feed 40 a year, in oats 170, and in potatoes 400. So the beef and mutton would have to be cut down. The Scottish farmer has long been in the habit of taking the plough round the farm, following his crops with temporary grasses. He worked a six-course rotation: oats, roots, oats or barley, then three years of grass. But even then, with 20 acres of arable, he might have as much as from 60 to 100 acres of permanent grass on his higher land. He was now asked to cut out one year of his rotation grasses in favour of a food crop, and to push his cattle farther up the hills away from the better permanent grass, which could then be ploughed.

And so the plough came to the grass, here, as in the south. Although, to the farmer's mind, much of it was ordinarily not worth ploughing, he was content to do it for the

counties where most of the arable cropping is carried out. To the south-west stretch the dairy counties of Ayr, Dumfries, Kirkcudbright and Wigtown, their climate softened by a Gulf-stream sea which warms their winding shores. To the north and north-west, in the Highlands and the islands —a considerable slice of the country—the land is mountain moorland, grazed by sake of the extra number of people he could feed. The first target was a stiff one, and it was not achieved without difficulty, for war brought its own special handicaps. The Scot springs from a long tradition of fighting men; he has a flair for battle, and the kilted Highland regiments have always been numbered among the crack troops of the British Army. Many thousands of young agricultural workers, therefore, already veteran Territorials, were called to the colours at the first Proclamation, and at one stroke the farmer lost a quantity of his skilled labour. Tractors, too, which might have compensated for that loss, were very scarce, and for the first two years there were practically no crawler types at all, though they were more urgently needed on the steep, tricky fields of Scotland than anywhere else.

But the farmers held out and increased their cultivations as best they could, until the shortage of labour and equipment was eased. Presently they got the tractors— 16,000 of them. They got 7,000 land girls to help to operate them, and nearly 300 blacksmiths, trained in oxy-acetylene welding, to provide maintenance centres in the scattered parishes. With rocky fields, steep hillsides, and many acres returning to rough moorland, it was not a straightforward job, but the area brought under the plough was well past the 1918 peak of the last war.

What was produced from this new land? Oats, of course. Oats that grow anywhere, in hard climate, on high or low land—oats for the porridge of north and south, and to swell the spare rations of the dairy herds. The production of this crop increased by 50 per cent. Oats are food for man and beast, and there are none better than those grown on Scottish soil. Potatoes, too, both for seed and table. Scotland is the nursery of the British potato, and its soil the source of many types long known and fancied by farmers and gardeners everywhere. The importance of the potato as a wartime food needs no further stressing, but it may not be generally known that the bulk of the vast

THE RIGS O' BARLEY. One of the old crops of Scotland, grown in the eastern counties; the base of good bread and malt.

new acreages planted in England during the war were sown with Scottish seed. For a very good reason too. If the English farmer plants his own or local seed for more than two years in succession, it runs the risk of contracting serious virus diseases. In Scotland, however, the cooler climate tends to keep the seed sound and healthy.

So there was unprecedented activity in the potato-fields of the north, more and more land being ploughed and planted. There could not be too much seed grown. You would find the machines out ridging the new fields early, for frost is a spiteful visitor right through the spring. By June and July you could see the red, purple and white flowers of the plant spreading in the soils round Loch Leven, in Cromarty, Black Isle, Aberdeen, Angus, Perth, and throughout the Lothians. And later the lifting and riddling of seed, the women and children packing them like thousands of golden eggs ready for shipment south. They were handled with special care, since they were the seed for the whole country—for the million-ton harvests to be produced on the English farms and the reclaimed Fens. These harvests would hardly have been possible but for the exertion and drive of the Scots seed-growers. Altogether they were exporting to England nearly half a million tons a year in seed alone —about four times the quantity exported before the war.

Other crops, both new and old, advanced up the sides of the eastern valleys, ousting the old grasses, and demonstrating to the farmers what the plough and a cartload of lime can do with land that once seemed old and tired. In Moray, Fife, the Lothians and elsewhere, the barley once grown for whisky was now grown in greater quantities for other uses. Enough flax fields arose to supply the needs of several new factories; and sugar-beet, that most unpopular plant, showed its broad leaf on many farms in spite of the farmer's natural inclinations against it. Wheat, a normal and satisfactory crop in a few favoured eastern areas, was springing

SEED FOR THE SOUTH. Scotland is the nursery of the British potato, and its soil the progenitor of many famous types long grown and fancied everywhere.

BLACK ANGUS BEEF STOCK.

LONG-HORNED HIGHLAND STEERS.

up on double the pre-war acreages. And round the Firth of Tay, where the climate is especially fair, the autumn fields were smothered with acres of crimson raspberries.

But none of these crops can be relied upon till harvested. Winter ploughing in Scotland is faced often with extreme difficulties, the year is long with labour, and harvests are tedious and slow, sometimes trailing late into November. As year by year the burden of harvests increased, the farmer's difficulties increased also; but, as in England, he received welcome help from volunteer workers. All round the Firth of Forth, and from Aberdeen to Berwick, they came from the smoke-belts of industrial Glasgow and the Clyde to live in camps and spend their holidays in the fields. Many schools closed down altogether in October to cope with the potato harvests. "Those schoolchildren", said a farmer, "have been lifting my potatoes right out of the jaws of winter." For when the frosts come in November, it is then too late.

You would see fewer livestock in the east then, but you would still see large numbers. Though hampered by loss of pasture and feeding-stuffs, the farmer made every effort to keep up the nucleus of his herds. For some were breeds of ancient fame, the disappearance of which would have been a disaster to the whole pastoral industry.

Scotland is famous for its beef breeds. Pedigree herds of Shorthorns and Aberdeen-Angus, many of world renown, are dotted throughout the arable areas of Scotland. Each spring, before the war, bulls of these breeds in the markets of Perth were fetching four-figure prices from buyers representing breeders in all parts of Europe and the Americas. The sleek, black, smaller-boned Aberdeen-Angus breed had its birth in the

AYRSHIRE DAIRY CALVES.

Scottish counties from which it has taken its name. These " blacks " are still being raised in their native valleys. Aberdeen store-cattle, fattened on local turnips and oat-straw said to possess a feeding quality unknown to the rest of Britain, still yield their fine beef long famed in the markets of Edinburgh and London. In the flat wind-swept fields of Caithness, in Easter-Ross, in Black Isle and Moray, and along the foothill fringes of the southern arable counties, cattle raised in Ireland are brought every year to graze on those pastures unfit for ploughing, or to feed on the oat-straw and roots provided by the cropping rotations.

Almost anywhere in Scotland you find sheep, and in the great hill grazing areas you are indeed more likely to see a sheep than a man. The east is still a great mutton-fattening area. Lean, hardy Blackface lambs and Cheviots, bred in the Western Highlands, come down to thrive on the sweeter lowland grasses in autumn and on the succulent turnips in winter. Cattle and sheep are an integral part of the Scottish arable system and so must be conserved as far as possible, for they bring a fertility to the soil which nothing else can replace, and they are the native strains upon which much of Scotland's wealth depends and will continue to depend in the future.

For the famous dairy herds of Scotland, we turn south-west to Dumfries, Kirkcud-bright, Wigtown and Ayr. This is not lush land, but it is warm and the rainfall is right for grass. A broken spine of granite mountain rises steeply through the middle of the area, a gaunt wall to catch the wind, and a home for sheep. On the green lower slopes that run towards the sea grow the pastures where the milk herds graze. Here you will find the small, wedge-like, red-and-white Ayrshire cows, which have been bred in these parts for generations, and on the poorer land shaggy, long-horned Highland and Galloway steers cropping the rough grass. From here, too, come the powerful Clydesdale draught horses, long-maned, feathery-hooved, and splendid in action.

This is a dramatic countryside, impressed throughout with signs and memories of old Border warfare. From the farms of Grey Galloway you can look out over the Solway Firth to the far blue hills of the Cumberland lake district. But they suffer from isolation and poor transport facilities. In bygone days the farmer had to turn the bulk of his milk into cheese, because that was the only way he could get it safely to market, a hundred miles away. The wartime demand for liquid milk forced him into a new life. Cheese-making on the farm finished, and his herds must be constantly improved. Moreover, the war proved a great stimulus to the eradication of tuberculosis. This district had been a pioneer in voluntary milk-recording and in the fight against tuberculosis, a consistent policy which resulted in a tremendous demand at high prices for native Ayrshire stock. But, among other changes, the dairy farmer had now to go out and plough for winter fodder, and that is something he had no need to do before.

Ploughing was not nearly so simple for him as for the dairy farmer in the south. His fields, though covered for the most part with good grass, are prone to sudden treacherous outcrops of rock. Throughout Grey Galloway the fields are split and torn by lumps of granite which give the landscape the look of an archipelago—a green sea broken by thousands of small jagged islands. And just as many lie invisibly submerged, covered only by a thin skin of turf. Yet between these rocks the soil goes deep and is good.

To plough such land without breaking a hundred plough-points called for an intimate local knowledge of each field. The ploughman had to guide his plough as delicately as a pilot bringing his ship to a rocky harbour. The Galloway farmer did this—ploughing in small patches, jumping and skirting the rock, an exhausting nightmarish job, but a job nobody else could do for him. Neither were tractors of any use to him here; it was a job between local man and horse. But it was worth doing all the same, and for the sake of winter milk and winter feeding—a far graver problem in Scotland than anywhere else—it had to be done.

The demand for milk gives rise to problems which continue all the year round; the pedigree farms of Ayr, Lanarkshire and Renfrew are conscious all the time of that vast industrial area around Glasgow of three million dependent workers, whose need cannot be wholly appeased in spite of the farmer's every effort.

From the pastures to the granite mountains which invade this whole dairy country, we again find sheep, always sheep. Sheep clinging precariously to the slopes, picking fastidiously among the sparse grasses, surviving still where little else could. It is here that a highly successful experiment in sheep migration was undertaken by the Department of Agriculture. Three years ago a certain area in the Northumberland hills was cleared for military purposes and a flock of 2,000 local Blackface sheep thereby rendered homeless. They were a good stock of ancient standing; in fact, the head shepherd had been with this particular fold since boyhood, and his father and grandfather before him. It seemed a pity to break them up and scatter them now. So they were adopted by the Scottish Department, loaded upon two special trains, carried 150 miles, and set down on a derelict moor above Castle Douglas.

It was a tricky venture. You do not move sheep at all easily; they have a special fondness for their familiar hills and grow obstinate or hysterical away from them. But this flock was soothed into its new home by five experienced shepherds who spent the first winter months leading them along the unfamiliar mountain runs. Nevertheless, when the snows came, heavy losses were expected, for the sheep would not know where to find shelter. But fortunately the weather was mild, and by the following spring the ewes were so weathered to the new surroundings that they settled down to a lambing which was highly successful. This experiment, an unusual one for Scotland, maintained unbroken a first-class flock and provided a steady flow of mutton and wool from a hill which was otherwise unused.

Away from the arable areas and the cities, away to the north-west, lie those 10,000 square miles of Scottish Highlands loosely described as "crofting and hill-sheep pasture". They stretch from Perth and Argyll, through Inverness, Ross, Sutherland, and include the Hebrides and the Shetland Isles: ranges of mountain, loch and forest, heather and bracken, rock and yellow grass. There is no other land like this in the world. And there are no other people quite like the crofting communities which live upon it.

There are perhaps no more than a couple of hundred thousand of them in all that scattered region; but they are a fixed and jealous population, bound to their crofts and their mode of life by the strongest ties of blood. They are among the most Celtic of all Celts and still retain the most apparent indications of their origin in looks, speech and music. Until the advent of the railways 70 years ago, they were wholly Gaelic-speaking, and the Gaelic remains still the natural tongue used by the old folks and by the children at their play.

Their crofting system is homogeneous and has a strong communal basis. The crofts are hereditary family small holdings, lying grouped together around the shores of the lochs and along the banks of the rivers and upon the saw-edged western and northern sea coasts. The holding will be only a few acres of tillable land, each marked with its steading and cottage and carrying its bare subsistence crop of oats or turnips. On the coast the crofter will divide his time between sea and land, part farmer, part fisherman.

These small holdings, grouped into communities or "townships", represent the

THE CLIPPING. Gathered from the hills in the late spring, the sheep are shorn, rebranded and turned loose again. Their wool is a modest though steady source of income.

THE GAELIC CROFTS lie along lochs and rivers, around the north-west sea coasts and out among the islands. Each crofter works his own small field and runs his stock on the communal grazings in the hills. Life here is hard, but self-sufficient.

greater part of the tillage area of the Highlands; they are small indeed, but above them hang the great mountains, and upon these the crofters have almost unlimited grazing for their livestock. Townships frequently run a communal fold on the hills, but in general each crofter has grazing rights attached to his holding; these are often based on the annual value of the croft and allow him to graze so many sheep or cattle of his own. The seasonal tasks of lambing, dipping and shearing of sheep are shared by the whole community. When the wool is clipped, it is spun and woven by the women and shrunk into fine tweeds. The crofters are a hardy lot and live a life with little margin for comfort, though plenty for happiness. Their whole existence has been devoted to the task of coaxing food from surroundings that are anything but bountiful. War did not make their lives any more austere or active than they normally were.

But war brought several peculiar influences to the Highlands, and many positive benefits. The hill-sheep industry, as distinct from the crofting settlements, had by 1939 reached the lowest ebb in its history. Hill-sheep farming proper consisted of huge folds of Blackface, Cheviot and Shetland sheep which stayed on the hills winter and summer alike, breeding there and foraging for themselves, providing wool for their owners, and lambs for fattening on lowland farms. But the popular taste for this slow-maturing mutton had decreased. The Highland cattle, too, were fast disappearing, and bracken was spreading like a plague up the glens, till the sheepfolds remaining could hardly find sufficient pasture to keep alive. A proper balance of sheep and cattle had to be restored to the hills, and a special type of ewe bred to meet modern conditions. To effect this, the Department of Agriculture backed up the hill breeder and supplied on hire first-class rams and bulls to father the necessary types.

Both for the sake of the Highlands and for the sake of our food-stores, the glens had to be grazed. The Department took over a quarter of a million acres of unused deer-forest, and grazed them with large flocks of sheep and hill-cattle in the care of crofter-shepherds and herdsmen. In 1942 five of these flocks produced a total of 165,000 lb. of mutton, 54,000 lb. of beef, and 73,000 lb. of wool from erstwhile forsaken glens.

These deer-forests are not, in fact, forests at all any more, for the old Caledonian forests which once covered them were destroyed by fire centuries ago. They are now those breath-taking craggy glens, spouting with ice-blue streams and hung round with mist and rain, the coloured counterfeit of which one may see framed upon a thousand parlour walls. For some years past, this type of country had been devoted to deer-stalking at the expense of sheep. The crofters themselves did not mind so much; it brought them certain advantages. They often earned good money as gillies and beaters; and, fortified by the old Gaelic saying, " It is no crime or shame to take a fish from the river, a tree from the forest, or a stag from the hill ", they saw to it that many a haunch of good venison found its way home to sweeten their frugal cooking-pots.

Deer-stalking, however, is not a wartime industry. Deer and sheep can live together; there is room for both in the right proportions. But by 1939 the protected deer population had so enormously increased that the land in many areas had grown too coarse for sheep altogether. So the official eye fell upon the deer, and thousands were condemned for the sake of their flesh (unrationed) and for their living space. Large expeditions were organised in which whole crofting townships took part, and the superfluous animals were tracked down and shot. The land was then re-sweetened for the sheep, the drains reopened, heather burnt, the bracken cut down. Innumerable sheep-runs were restored in this way.

By everything in their power the Government encouraged the hill farmers and the crofting townships to restore balanced grazing to the glens and to exploit once more the productivity of those vast areas in the most natural way. This may be no land for the plough, but it breeds good meat.

War did not solve the Scottish agricultural problem, or make life any simpler for the Scottish farmer. But at least he could see before him a promise of great developments once the restrictions of war were lifted.

He ploughed more widely than for a hundred years, fertilising, liming and cleaning his land, planning his livestock breeding and getting accustomed to the temporary heaven of an ordered market. He had before him the example of a Government organisation fixing prices ; hiring him labour and machinery ; selling him seed ; sharing the cost of his ploughing, draining and land reclamations ; supplying him with stud horses, bulls and rams for his stock ; analysing his soils, giving him expert advice on all his problems, animal, vegetable and mineral ; and, finally, buying his produce from him, or at least guaranteeing its sale. He felt his scope increasing in spite of wartime limitations, for he brought Scotland nearer to self-sufficiency in food than she had been for three-quarters of a century, and that in the face of war and a vastly increased population. With this experience behind him, and the abundant facilities of peace before him, what new heart and fertility may he not bring to his rugged but magnificent countryside?

9

Ulster: a country of small holdings

IN NORTHERN Ireland, the story of the wartime farm was very similar to that of Great Britain, differing in detail, but much the same in achievement. There were similar shortages in tools and labour, a similar drive to plough up old grass, the same efforts to produce more from less. But there was a difference both in method and in the type of crops grown, for the Ulsterman farms a different sort of land and he farms it in his own way.

Northern Ireland is barely the size of Wales. It pivots around the wide fresh waters of Lough Neagh, and in appearance suggests Cumberland or Cornwall. The mountains of Londonderry, Antrim and Down are rough, gorse-covered, and cropped with volcanic rock. The land is loaded with historical relics, ruined castles and Celtic crosses, old battle-grounds, peat bogs, swift-running rivers, and gentle valleys of good soil. It is a country farmed chiefly by small peasant proprietors, a land of small holdings worked mainly by the family itself, and there is consequently a smaller proportion of outside labour. There are no landlords here and no big estates; the family works and lives on its own land, and 70 per cent. of all holdings are less than 30 acres in extent. There are, in fact, a greater number of farms in the six counties of Ulster than in the whole of Scotland.

Ulstermen have always inclined to mixed farming, having many irons in the fire at once and working the whole agricultural gamut—cattle, sheep, pigs, poultry, milk, grass and cultivations. In this way they were well insured against disaster. If crops failed, they lived on their livestock; if eggs or bacon slumped, then they had other alternatives. This reluctance to specialise stood them in good stead through the great depression between the wars, which hit them less severely than it did many of the single-department farms of Britain.

Moreover, when war brought its demand for increased cultivations, the Ulster farmer was pretty well prepared. He merely had to concentrate on the arable side of his business, and it caused him little upheaval. For many years he had been farming his few fields on a system of planned rotations which only came to the South of England with the wartime introduction of the ley system. His method was to plough up a field of grass and sow it first with oats, then flax, then roots (potatoes, mangolds or turnips), then oats again undersown with grass. The grass following the final crop of oats was cut for hay the first year and subsequently grazed. After grazing, a period of restoration lasting from two to five years, the plough returned to the grass and the cycle began again. This method catered

PEASANT PROPRIETOR. His cottage lies snug beneath the hills of Antrim. He keeps a few pigs, some sheep and poultry, has a field of barley and a cow in milk.

both for crops and cattle. A proportion of the rye-grass grown was saved for seed, for three-quarters of the world's rye-grass seed comes from Northern Ireland. Thanks to the rhythm of this rotation, already well established, the farmer was able to move easily into war production, to increase his output of crops and to maintain his head of cattle.

The control and organisation of the food front were in the hands of Ulster's Ministry of Agriculture, which operated its own variations of the English system. The pre-war County Agricultural Committees were taken over and equipped as Divisional Headquarters to direct campaigns and strategy, and their staffs were well stiffened by additional technicians and experts. The farmer worked in direct contact with the Government; it sold him his seed, fertilisers, machinery; it bought nearly all his produce. This represents millions of paper transactions and much heavy staff work. But the farmers' personal leader, a civil servant, was the Area Tillage Officer, an experienced man of the soil who went out to the fields, advising, directing, assisting, and keeping constant watch on all operations.

HARDY CROPS for a hard climate. *Above*, oat-harvest in Glenariff; *below*, the " retting " of flax—the first stage in the making of linen.

The first signal for general action came with the Government's Tillage Order, issued at the outbreak of war. It required every farmer without exception to plough up an additional percentage of his land. The farmer could decide for himself which fields to plough, but the increase was compulsory, and it spared no one. Consequently, each did his share without question and jumped heavily upon his neighbour if he suspected any backsliding. As a result of this Order, the land under crops in Ulster had increased by over 80 per cent. since 1939.

For a country of small farmers this was a considerable achievement, and it was carried out on top of many other commitments such as stock-raising and poultry-keeping. The labour problem was as acute here as in England, but the farmer was helped financially by a number of Government schemes. Tractors and other farm tools, Lend-Lease or home-produced, were all controlled by the Government and issued only to those in greatest need. Farmers could buy their tractors on a special hire-purchase system. Many holdings were, of course, too small to possess much modern equipment, but they frequently needed its help all the same. So the Ministry of Agriculture arranged for tractor-purchasers to plough up their neighbours' fields at fixed rates of pay. This helped the owner to meet the cost of his machine and spread the benefit of it over a much wider area. By certain land improvement schemes farmers were assisted in bringing much second-class land back into production. The Government bore a large proportion of the cost of the lime and basic-slag needed to revitalise sour fields. It paid 50 per cent. of the cost of operations aimed at reclaiming bogged land and flooded areas—cleaning streams and ditches, opening hill drains, and clearing gorse from the higher lands. It also paid a subsidy to hill-sheep farmers, to help them maintain their flocks and to offset the decreased demand for store-sheep resulting from the plough-up of lowland grass.

In the building of flax-dams, so important to the flax industry, the Government bore the whole cost. Flax was the most vital wartime crop grown in Ulster. It was not a food crop, but an essential raw material used particularly in the manufacture of aircraft fabrics, flying suits, parachute harness, machine-gun belts, and so on. The Belfast linen trade, once largely dependent upon imported fibre from Latvia, Lithuania and Estonia, was now wholly supplied from local produce. Although in Great Britain we had begun to grow flax in large quantities, Ulster farmers had a much longer experience of it and were producing more than twice as much as England and Wales put together. Ulster farmers process their own flax, producing the retted fibre which is the only material that can be used where great strength and reliability are essential. The green flax, straw-grown and sold by English farmers, was used for coarser fabrics and this released as much of the Ulster flax as possible for more vital requirements. Flax growing and scutching is part of the Ulster countryside; flax fields give to it that particular blue-green colour which is all its own, and the streams and rivers are strung out with flax-dams and scutch-mills, some the overgrown relics of older days but the majority newly built to cope with the needs of war.

Care of the flax crop demands experience and a deal of personal attention. The plant must be pulled by hand, delicately and gently. It is true that machines exist for this operation, but they are few, and the crop is so frequently battered and tangled by rough weather that only the hand can deal with it effectively. After pulling his flax, the farmer still has much to do. First he carries it to the dam, going in up to his thighs to spread it out on the stagnant waters. Here it remains from seven to twelve days while the gums and woody matters soak and rot away. When sufficiently " retted ", it is removed from the dams and laid out on the fields to dry. After gathering,

ULSTER PASTORAL. County Down sheep, long-fleeced and black of face, browse in their morning shadows on the slopes of Lurigethan.

the flax is built into " barts " where a final drying takes place. It then goes to the scutch mill, where it is scutched over rapidly revolving handles which separate the pure flax fibre from the woody material. Finally the finished fibre returns to the farmer, who in war time carried it to the flax market and sold it to the Ministry of Supply at a fixed price. During all these operations the farmer never loses sight of his flax; from the field to the Ministry of Supply its quality remains his constant care and responsibility. Since 1939 flax acreages were increased four and a half times.

The chief food crops coming from the new-ploughed lands were oats, potatoes, cattle fodder, and barley. In County Down a certain quantity of wheat was grown, but on the whole the climate is too treacherous for wheat. Oats and barley were the toughest crops, and oats was the crop produced in the greatest quantities. But at harvest time the flax crop, being the most sensitive of all, must be taken first, while the others await their turn. Because of this, and a spell of vile weather, the 1942 and 1943 harvests were long and bitter struggles. Those two seasons were the worst Ulster has known in 50 years. Much of the corn lay flat and could not be dealt with by the usual binders, and the harvests dragged on right into November. But thanks to the help of soldiers, British and American, and of schoolboys and students, the harvest was saved.

As for cattle, here Northern Ireland is in a particularly strong position. Her dairy herds increased in spite of the shortage of feeding-stuffs. On an average there were $4\frac{1}{2}$ cows to every holding; milk was un-rationed and its consumption increased by over 100 per cent. Free milk schemes for children and mothers operated as in England; much surplus milk was evaporated and tinned for the Forces, and in the winter thousands of gallons were shipped daily to Scotland. The pig population, alas, was cut to a third;

ULSTER POULTRY. From hens like these came sufficient eggs each year to supply Greater London with its wartime ration.

one looked in vain for the magnificent Derry hams of old. But the number of sheep and bullocks raised on the hills made possible not only the supplying of all home needs but the exportation to Britain of millions of pounds' worth of meat each year—a noble contribution from so small a country.

The purchasing, slaughtering and disposal of all livestock were carried out by the Ministry of Agriculture, acting as agent for the Ministry of Food. There were no middlemen in Northern Ireland. Every cow, bullock, sheep or pig, from every holding in the country, was sold direct to the Ministry, graded by its own officials, killed in its own abattoirs, and sent direct to the butchers' shops or on hoof straight to the docks for shipment to England. The efficiency of this method saved the time wasted in individual bargaining, guaranteed the quality of the meat, and made full use of the by-products which are so often lost in the casual market. Government cleaning centres were built to deal with all hides, skins and offals; they set up their own tanneries and triperies, and even salvaged certain animal glands for medicinal purposes.

The maintenance and development of herds in Northern Ireland confounded the experts. Ulster's experience showed that the more land there is under cultivation, the bigger the head of stock that can be maintained. During the pre-war agricultural depression, when the price of grain was falling throughout the world, sheep, poultry and pigs were the salvation of Northern Ireland. And during the war poultry remained its strongest arm. The poultry population increased steadily there since the outbreak of war. The Ulster hen had long since ceased to be just a farmyard animal. It was the biggest single item in the farmer's income, housed and looked after with all the care and attention such an important contributor to the income deserved. It was in the fields, not in the buildings of the farmyards, that Ulster's poultry flocks were to be seen.

For some years before the war Ulster had a national scheme for the grading, packing and distribution of eggs, which worked so efficiently that it received world-wide recognition and was even adopted by Northern Ireland's competitors, the Danes. Eggs were sold by weight, not by number; a chain of packing stations throughout the country took them straight from the hands of the farmer and passed them on to the ports and cities, maintaining a constant flow so that stocks were never held up and always arrived in the best possible condition. In the war this machinery worked double time.

Eggs have never been short in Ulster; had the people chosen to keep their produce to themselves, we might well have envied them. In 1941, however, they submitted to a rationing scheme in every way as severe as our own. It transformed their country overnight from one in which you could buy eggs by the basketful to one where they were almost as scarce as oranges. Although they were producing greater quantities than ever before, the civilian got no more than his bare allotment of two to three a month. The entire surplus, and it was considerable, was shipped to Britain. We got 80 per cent. of all the eggs they produced—enough, in fact, to supply the whole of Greater London throughout the year. Such a gift, in those days, was a sacrifice indeed.

The small farmers of Ulster made a remarkable showing in the war; they took its problems in their stride, and turned their versatility to the best account. In spite of a broken, rocky country, treacherous weather and all the usual shortages, they stepped up production on all sides; their tiny holdings worked to a common plan, and in flax, meat, milk and eggs made prodigious contributions to the war stocks of the nation.

10

New life on the land

LAND is a pretty good mirror of man's state of mind. It reflects his outlook, his way of life, his standard of civilisation. A countryside of weeds and broken hedges will point surely to the demoralisation of the community living upon it, just as well-ordered cultivations will show its self-confidence and power.

The farmer must be judged, not by the look of him, nor by the clothes he wears, but by the way he treats his land and keeps his stock. The British farmer speaks well for Britain. Living as he does, he knows very well just what Britain means. But he expresses what he knows in farming, not in words. As a member of society he is unemotional, unromantic, canny, and very independent; he cares little for the sort of figure he cuts in the world. He dresses to work the land and fight the weather, and his mind is on that only. Looking at the land, after six years of war, you knew what the farmer was thinking; you knew that he believed in himself and in this country. And you felt that this country was going to survive.

The farmer was not always so sure of himself, however, nor had he every encouragement to be. Many farmers, of course, were forward-looking and forward-moving even before the war, but there were others whom circumstances had driven into comparative isolation, so that they seemed caught in the backwash of an older century. It appeared to them that the only thing they could trust was the land; it was something that did not change. Consequently they stuck to it and did not change either.

Most farmers took the new wartime life in their stride, but the men of this lost minority were resurrected by it. They had at last been given the opportunity to come out into the open, to farm expansively on scientific lines, to double and treble their output. They wanted nothing better than to be able to afford to do so—a privilege long denied them. As we mobilised for total war, therefore, they discovered that their skill was, after all, still one of our most powerful weapons, and they left the old world overnight and lost no time in catching up with the new.

The difference between the pre-war and wartime farmer was historical and significant. It could best be observed in the latter's new communal spirit, his enthusiasm for the sharing of experience, and his desire to learn through and with his neighbours. New crops and methods, continually changing with the sweep of war, brought farmers together everywhere to compare notes and exchange theories. They co-operated in every possible way—organising discussion groups, meetings, lectures, demonstrations,

BRUSHING UP ON THE PLOUGH. Ploughing has always been a skilled man's work; the tractor has introduced a new technique into it. County Committee demonstrations bring home the best methods to every farmer.

determined not only to pick one another's brains, but to spread the gospel of their experience among their neighbours.

"Neighbours' Day", for instance, became one of the most popular and direct methods of contact. A farmer, taking his prestige in his hands, would invite his neighbours to call on him, say, the following Sunday morning. They arrive, walk round his sheds and fields, observe, discuss, and ask questions. It may be something of an ordeal for the host; he knows that, if he has made mistakes, no sort of excuses can camouflage them now. If he has been particularly successful with his crops or livestock, then his neighbours will learn by it, and his pride will have its day. And if anything has failed, someone will tell him why, and he, in turn, will benefit by the advice.

In a less direct but equally important way, discussion circles served the same purpose. Meeting together in pub., schoolroom, or farmhouse kitchen, countrymen spent many a long winter's evening holding inquests on the year's achievements, swapping theories and opinions, airing old feuds, and planning the next spring sowings. Such gatherings were a real tribute to the War Agricultural Committees and to the energy of their hard-worked staffs. Many of these men attended discussion groups three or four times a week, and this after a full day's work in the fields. These discussions were a new thing in village life. Attend one, and you would get a pretty good idea of the spirit among farmers.

Sitting round with thumbs in waistcoats, puffing at battered pipes, they are then at their most vocal. There are old men and young men, their faces polished by wind and rain, their hands as brown and gnarled as walnuts, all weighing their words with special care. There may be some present who have not met in the same room since boyhood, sworn enemies perhaps, and they will not refrain from taking a sly dig at each other—though it is just as possible here they will make it up. Then there is the old farmer whose dialect is shrouded with archaic allusions, tags and formulæ; who knows the history of every field in the neighbourhood for generations back, what was grown there, when and by whom, why it thrived or why it failed, and says what he thinks of the damn-fool attempts of some of the younger men. He comes in for a certain amount of leg-pulling; but everyone listens to him, for he is the land—not just any land, but this particular stretch of land folded in this particular valley which, like him, has moods, tricks and wisdom to be found nowhere else.

As these men talk, you find their phrases short, to the point, falling on the ear like proverbs or blank-verse quotations, exceedingly simple and memorable:

" Never touch that field after Christmas."
" Ground's too cold."
" Cold as my grandmother."

" One acre down here's worth three up North Common."

" Must have good pasture close home, Bert.
　I work mine one and one; mow it, then graze it."

" Broadcast kale and you'll beat the wireworm,
　that's what my father used to say."
" I think with your father, Jack—they can follow a drilled row—they get lost when it's broadcast."
" I sowed maize in my ten acre—the rooks took it—
　I said, Go on, make a good job of it—
　and they did.
　Then I sowed flax—it thrived."

They talk with new words, too, though delivered with the same sure sense of truth:

" What d'you balance your mangolds with? "
" Super and sulphate of ammonia."
" Two hundredweight sulphate to four of super."
" Lime and ammonia won't mix."
" Any antidote for fly? "
" Derri-dust."

And so on. Out of all this the farmers built up among themselves a sound practical relationship. They may have had to give away something in the process, some private piece of knowledge guarded jealously for years perhaps; but, for each one surrendered to the common pool, they picked up many that were new to them. The real significance of this type of discussion is that it was based on local experience, soil conditions, and personalities which vary from parish to parish and for which no outside educational effort could possibly cater so well.

Young Farmers' Clubs, too, had a far-reaching effect upon farmers in general. For the older men are more readily influenced by their sons than they might be by strangers, and the development of Y.F.C.s throughout the country was responsible for the spreading of new technical ideas which went much further than the young farmers themselves. Y.F.C.s were started in this country over 20 years ago, but they had never flourished so well as during the war. They were as important socially as educationally, for they were open to everyone. Members' ages varied generally from 14 to 30, though this was no fixed rule (one of the staunchest supporters of a Cotswold Y.F.C. was an old cowman of 85). The clubs were democratic, and elected their own administrative committees on which young farmers, workers and land girls shared an equal responsibility.

The educational facilities offered the whole farming community had, in fact, never been more ambitious. Over a million and a half " Growmore " leaflets were distributed by the Ministry of Agriculture, with concise notes on every conceivable problem. Wherever farmers congregated you would find its experts plugging practical demonstrations of one sort and another. A practical demonstration, particularly if it is held on a neighbour's farm, means more to the farmer than any amount of literature or expert talk, no matter how knowledgeable. He is more impressed by the modest achievements of a neighbour than by the most sensational tales of a stranger's successes. And on the grounds of " Well, if old Tom can do it I reckon I can ", is more likely to emulate his example.

You would find the farmer and the farm worker attending such demonstrations in a spirit of quiet, deliberate inquiry. Conscious that theirs was one of the most complex industries that exists, they were out to pick up anything that would add to their skill and technical scope. That is the main difference between the modern farm worker and the old-time labourer—his conscious pride in himself as a first-class technician. Not so many years ago he was the most ill-paid and overworked individual in the country. He seldom earned more than 30s. a week, drew neither dole nor sick benefit, and got no holidays. Since the war he has begun to receive his due ; his wages have risen to a more reasonable level, his present minimum of 70s. a week comparing with 48s. in 1941. He is paid overtime and has a week's holiday with pay ; and he had already begun to enjoy State protection in sickness and unemployment for several years before the war. Moreover, the nation knows his real need for better living accommodation.

The growth of his trade unions has been responsible for winning him increasing benefits and a recognition in keeping with his extreme importance to the community. The national conscience, expressing itself in such constructive bodies as the Agricultural Wages Boards, had begun to see to it that he received a better and more just reward for his labours. This attempt to raise his standard of living is of the utmost importance to the future of agriculture. For a sense of absolute fatalism about his conditions had for years been developing in the worker's mind, so that he accepted them as the inevitable evils of his profession. The only protest many allowed themselves was the negative one of throwing up their jobs completely and moving away to the towns. This movement became so general that at the outbreak of war there were fewer skilled men on the land, relatively, than ever before. And of those remaining, 50,000 were absorbed by the armed forces within the first two years. With 6,000,000 additional acres to work and cultivate, the farmer was faced with a very grave labour shortage indeed. Although this shortage was never overcome, it was eased to some extent by auxiliary labour, by the return to the land of many thousands of townspeople.

The most important contribution to this movement was, of course, the one made by the Women's Land Army. This young, sun-tanned, green-sweatered, cord-breeched army of one-time shop girls, typists, mannequins, mill girls, hairdressers, parlourmaids, was throughout the war in the forefront of the battle. They figured in every ordeal and triumph wartime farming has had to offer. They filled the early gaps in manpower. They drove their ploughs on the winter pastures of 1939; they were the spearhead tractor-drivers of countless tough reclamation jobs; they went through all the campaigns of winter and summer, struggling with storm and mud, milking, sowing, harvesting, threshing, living in wild unfamiliar parts of the country, among strangers and stranger ways, often completely cut off from the world in which they had been brought up. An army of girls, tackling a life which many a man would find exhausting, suffering from callouses, chilblains, and aching bones, but buoyed up all the time by an extraordinary curiosity and devotion for those tasks which seemed to be, somehow, so much more vital than the aimless jobs of city and factory to which so many had been accustomed.

The W.L.A. was set up some months before the war. The first thousand members went straight on to the land, though most recruits after that received preliminary training for their particular jobs. But the land girl throve on direct experience. Her worth to her employer may be judged from the labour conditions guaranteed her. Each girl was engaged by contract which guaranteed her regular employment throughout the year, a minimum of one week's holiday with pay and no deductions for sickness or wet weather. Those under 18 received a minimum weekly wage of 18s. over and above their keep, and those over 18 got a minimum of 22s. 6d. They could always be sure of this much in their pockets, and any overtime increased it. Such wartime conditions were the legal obligation of the employer; what is more important, they constituted a reasonable precedent which may well influence the future conditions of agricultural labour, both male and female.

The W.L.A. officers and County Committees, assisted by 4,000 voluntary local helpers, dealt with the placing, billeting, equipping and welfare of the girls, this welfare work including the organisation of clubs, training, and proficiency tests.

The coming of the land girl, in the early days, was a source of some perplexity to the farmer. To his wife also. Neither was quite sure how it would work out. Compared with the ruddy, strong-limbed village lasses, these paler, streamlined, town-bred girls seemed much too fragile for the rigours of outdoor work. But farmers' wives and country women generally, in spite of the additional burdens placed upon them by war, gave the girls a great welcome, housing and looking after them, and easing for them the unfamiliarity of their surroundings. And the girls, fragile as they may have seemed, soon surprised everyone, not only by their fresh enthusiasm but by their almost obstinate fanaticism, which refused to allow any job, mucky or otherwise, to get the better of them. There was nothing they would not attempt. Some were particularly suited to the new life, anyway: Lancashire girls, for instance, accustomed as they were to an industrial environment, adapted themselves readily to the subtleties of farm mechanisation. As a Wiltshire farmer said of another type of girl: " Of course I was a bit doubtful at first, especially when I found I was let in for a couple of actresses off the stage. But let me say this—if actresses can make such good land girls, let me have actresses again !"

Apart from their contribution to manpower these 90,000 land girls formed a very strong bridge between town and country. Many married farmers and farm workers, and settled down altogether to the new life, while hundreds of others were taking up agriculture as a serious study and intended to make a permanent career of it.

Many other types of auxiliary labour beside the W.L.A. were brought in to fill up the gaps, particularly for the heavier work of ditch-digging, bush-clearing and so on. They were strange types too, some of them—men who, but for the war, might never have known what a British farm looked like—Italian prisoners of war, and aliens and refugees from all parts of Europe.

Yet, in spite of all these, there were certain times of the year, still, when the labour situation remained exceedingly critical. It is easier to sow a crop than to gather it. Yet farmers continued to sow thousands of additional acres each year in the faith that, once sown, the means would be found to harvest them. They had no idea at all where those means would be found, but in one way and another they were found. In the first years, for instance, soldiers based in the heart of the country came to the aid of farmers in whole battalions.

Later, of course, the armed forces had other business on hand and much less of this kind of help was available to the farmer. But other armies took their place, armies of unskilled but enthusiastic civilians. For, as in olden times of common peril, when the need for action would be felt instinctively by the whole community, so in this war, men, women and children in their thousands came out from the towns and villages in time to help get in the harvests. Each year the sowings were greater, the harvests heavier than the last. Each year it seemed inevitable that some would be left to rot. Yet, as they ripened and stood over the countryside, heavy, golden and ungathered, an indefinable sense of urgency seemed to communicate itself to the townsman, so that at the last moment he was always there to lend his strength to the farmer.

Schoolchildren gave great help everywhere. Those from the country were used to seasonal work on the land. But great numbers came from town and city too, from public and council schools alike—boys and girls to whom the produce of the farm had never been much more than items on a shopping list. It was a great adventure for

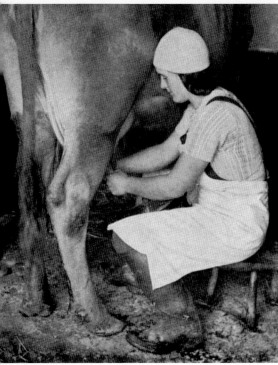

LAND ARMY. They came from shops, offices, beauty-parlours, dancing schools. They went wherever they were needed, did any job that would replace a man. They have known six winters and six summers on the land, have gathered six wartime harvests.

POTATO RACE. In spare time and holidays they have done good work, sowing, reaping, picking up pocket-money in their stride. Farming, for them, has been a worth-while game.

them; they came in loud, excited gangs, bursting with curiosity for the new world. They lived under canvas or in specially requisitioned premises, and worked out their holidays among the fields. They helped with every kind of job: the boys potato-planting and lifting, tractor-driving, harvesting, flax-pulling, root-hoeing and singling; the girls potato-planting and lifting too, weeding, pea-picking, fruit-picking, flax-pulling—their neat swift fingers unrivalled at such labour. In 1942 the boys and girls of Britain worked nearly 10,000,000 hours in the fields. They did much more in 1943, and arrangements were made for them all over the country to meet the still larger harvests to come. Surrounded by the clamour of war, they appreciated very well the reality of what they were doing; they knew their work was important and, of course, they were being paid for it.

As for the adults—village women, like their children, are still close to the tradition of communal farm work. The days are not so far distant when they were expected, as a matter of course, to join their menfolk in the fields at certain times of the year. True, the custom had begun to die out, at least till war came along. The women went back to the land; Women's Institutes organised all kinds of part-time labour for them, and members volunteered for service whenever their housework would allow them. Many who could spare a couple of days a week volunteered as relief milkers to give cowmen and land girls an occasional day off. Others, organised in the manner of midwives, helped to attend the calving of cattle and so to relieve the farmer of yet another of his anxieties.

But, during the critical weeks from June to October, there was never enough labour either in the yards or the villages to cope with the vastly increased cultivations. Labour was needed throughout all those weeks, right up to the harvest. If, during the war years, volunteers had not come to his assistance, not even the farmer could tell you what setbacks might have occurred. The movement which brought the town workers to the fields is a significant one. It proves that the hereditary link with the land is still unbroken, even among the most urbanised communities. Perhaps too that sense of food, which always becomes more acute in war time, persuaded some to look to the fields to see how things were going. Anyway, Land Clubs for volunteer labour began to spring up in towns and cities all over Britain. Factory hands sent deputations to the W.A.E.C.s, offering their spare time and demanding to be put to work.

Civil servants, bank clerks, solicitors, typists, shop girls, employees and bosses alike, joined with them. Harvesting, suddenly, became everybody's business.

The Agricultural Departments and the Ministry of Labour threw themselves wholeheartedly into the task of organising this vast and willing force. Camps and hostels were erected and placed strategically throughout the arable districts. Village halls and schools were requisitioned. They were fully equipped with bunks, food kitchens, and maintenance staffs. Then they were thrown open to everyone: here you could come and spend your holidays; you were promised plenty of fresh air, toil and sweat without the tears, a temporary land job of first-class importance, and a round payment of a shilling an hour as well—age and experience no object.

To these camps the workers came. They

IN THE SUN-RIPE PEA-FIELD, they work and gossip—village women, with their children, by age-old custom the farmer's seasonal help.

HOLIDAY TASK. City workers came in their thousands; some tore off their shirts, and all worked with immense gusto. The harvest had become everybody's business.

came in their thousands; they forgot Blackpool and Clacton; they forgot coal-dust, oil-waste, buzzing lathes, blackout-boards; they tore off their shirts, took a good swig of country air, and plunged into the corn and cabbage as if for a salt-sea bathe. Many were awkward, stiff-jointed with their tools, but they worked with immense gusto. And they got a smell of the land; they saw what a crop looked like, what it was like to handle, and what labour went into the raising of it. They knew the satisfaction of a good day in the harvest fields, with dust and chaff stuck to your sweating arms, hours in the hot sun and the exhilarating thirst it gives you, and best of all the solid appetite you carry away with you in the evening, the job well done, the corn ready stacked in the barn.

Perhaps they were never on the job longer than a week at a time; but when they left, others took their places throughout the summer. Thousands of these helpers handled the bumper harvests of 1943 and 1944.

When at last they got back to their desks, counters and work-benches, stiff as they were, blistered, burnt red with the sun, their food tasted better to them, the countryside had more significance, and the farmer's life—his problems and the tasks with which he is faced—seemed more real to them, less of a fable, because they had shared it with him.

Surely the gulf that has hitherto existed between the two communities, town and country, could be bridged in no better way than this?

WAR RECORD: A SUMMING UP

LAND: Since 1939 6,500,000 new acres have been ploughed up.

LABOUR: The land has lost 98,000 skilled men. But 117,000 women have replaced them.

LIVESTOCK: Between 1939 and 1944 milking cows increased by 300,000; other cattle by 400,000. But there were 6,300,000 less sheep, 2,500,000 less pigs, 19,200,000 less poultry.

HARVESTS:

	1934-38 average Tons	1943-44 Tons	% Increase
WHEAT	1,651,000	3,449,000	109
BARLEY	765,000	1,641,000	115
OATS	1,940,000	3,059,000	58
POTATOES	4,873,000	9,822,000	102
SUGAR BEET	2,741,000	3,760,000	37
VEGETABLES	2,384,000	3,197,000	34
FRUIT	455,000	705,000	55

NOTE.—*The figures above are drawn from "Statistics Relating to the War Effort of the United Kingdom", a Government White Paper published in November 1944.*